Modern Birkhäuser Classics

T0235854

Many of the original research and survey monographs in pure and applied mathematics published by Birkhäuser in recent decades have been groundbreaking and have come to be regarded as foundational to the subject. Through the MBC Series, a select number of these modern classics, entirely uncorrected, are being re-released in paperback (and as ebooks) to ensure that these treasures remain accessible to new generations of students, scholars, and researchers.

For Milada

Complex Analysis

Fundamentals of the Classical Theory of Functions

John Stalker

Reprint of the 1998 Edition

Birkhäuser
Boston • Basel • Berlin

John Stalker
School of Mathematics
Trinity College, Dublin
Dublin, 2
Ireland
stalker@maths.tcd.ie

Originally published as a standalone

ISBN 978-0-8176-4918-0   e-ISBN 978-0-8176-4919-7
DOI 10.1007/978-0-8176-4919-7
Springer Dordrecht Heidelberg London New York

Library of Congress Control Number: 2009931686

Mathematics Subject Classification (2000): 30-xx, 33-xx

© Birkhäuser Boston, a part of Springer Science+Business Media, LLC 2009
All rights reserved. This work may not be translated or copied in whole or in part without the written permission of the publisher (Birkhäuser Boston, c/o Springer Science+Business Media, LLC, 233 Spring Street, New York, NY 10013, USA), except for brief excerpts in connection with reviews or scholarly analysis. Use in connection with any form of information storage and retrieval, electronic adaptation, computer software, or by similar or dissimilar methodology now known or hereafter developed is forbidden.
The use in this publication of trade names, trademarks, service marks, and similar terms, even if they are not identified as such, is not to be taken as an expression of opinion as to whether or not they are subject to proprietary rights.

Printed on acid-free paper

Birkhäuser Boston is a part of Springer Science+Business Media (www.birkhauser.com)

John Stalker

Complex Analysis
Fundamentals of the Classical
Theory of Functions

Birkhäuser
Boston • Basel • Berlin

John Stalker
Department of Mathematics
Princeton University
Princeton, New Jersey 08544

Library of Congress Cataloging-in-Publication Data

John Stalker
  Complex analysis : fundamentals of the classical theory of
functions / John Stalker.
          p.   cm.
  Includes bibliographical references and index.
  ISBN 0-8176-4038-X (alk. paper). -- ISBN 3-7643-4038-X (alk.
paper)
  1. Functions of complex variables.   2. Mathematical analysis.
I. Title.
  QA331.7S73 1998   97-52126
  515'.93--21   CIP
AMS codes: 30, 33

Printed on acid-free paper
© 1998 John Stalker   *Birkhäuser* ®

Copyright is not claimed for works of U.S. Government employees.
All rights reserved. No part of this publication may be reproduced, stored in a retrieval
system, or transmitted, in any form or by any means, electronic, mechanical, photocopy-
ing, recording, or otherwise, without prior permission of the copyright owner.

Permission to photocopy for internal or personal use of specific clients is granted by
Birkhäuser Boston for libraries and other users registered with the Copyright Clearance
Center (CCC), provided that the base fee of $6.00 per copy, plus $0.20 per page is paid
directly to CCC, 222 Rosewood Drive, Danvers, MA 01923, U.S.A. Special requests
should be addressed directly to Birkhäuser Boston, 675 Massachusetts Avenue, Cambridge, MA 02139, U.S.A.

ISBN 0-8176-4038-X
ISBN 3-7643-4038-X
Typeset by the author in Latex
Printed and bound by Quinn-Woodbine, Woodbine, NJ
Printed in the United States of America

9 8 7 6 5 4 3 2 1

Contents

Preface

All modern introductions to complex analysis follow, more or less explicitly, the pattern laid down in Whittaker and Watson [75]. In "part I" we find the foundational material, the basic definitions and theorems. In "part II" we find the examples and applications. Slowly we begin to understand why we read part I. Historically this is an anachronism. Pedagogically it is a disaster. Part II in fact predates part I, so clearly it can be taught first. Why should the student have to wade through hundreds of pages before finding out what the subject is good for?

In teaching complex analysis this way, we risk more than just boredom. Beginning with a series of unmotivated definitions gives a misleading impression of complex analysis in particular and of mathematics in general. The classical theory of analytic functions did not arise from the idle speculation of bored mathematicians on the possible consequences of an arbitrary set of definitions; it was the natural, even inevitable, consequence of the practical need to answer questions about specific examples. In standard texts, after hundreds of pages of theorems about generic analytic functions with only the rational and trigonometric functions as examples, students inevitably begin to believe that the purpose of complex analysis is to produce more such theorems. We require introductory complex analysis courses of our undergraduates and graduates because it is useful both within mathematics and beyond. Why then do our textbooks create the opposite impression?

Ideally part II would repair some of the damage done during part I. Unfortunately the demands of the academic calendar and the ambient mathematical culture have slowly eviscerated part II. Students now are lucky to see even the most elementary properties of the Gamma function and the Weierstrass P function. The fact that these functions are important beyond the confines of complex analysis is scarcely mentioned.

What is needed is a change in emphasis and timing. Part II needs to expand again, though perhaps not to the size it had in Whittaker and Watson's day. Part I needs to contract correspondingly. More importantly though, as much as possible of part II must be presented *before* part I, so that what remains of part I is seen to be genuinely useful.

The paragraphs above describe how I think the theory of analytic functions should be presented. What follows is my attempt to implement these ideas.

The intended audience for this book is anyone who has taken a calculus course, who knows or is willing to believe the elementary theorems of real analysis given in the appendix, and who wants to learn the classical theory of analytic functions. The pace is rather fast. In compensation the book is quite short. Classical complex analysis is based on a few, very powerful, ideas. They can be presented and illustrated quite quickly, if one avoids the many possible entertaining detours along the way.

The book has been kept as self-contained as possible. The prerequisites are minimal. All the real analysis required is given in the appendix. This has not entailed any dilution of the material covered. Indeed the book goes considerably beyond the standard first-year graduate syllabus in several directions.

Outline

This book is divided into three chapters of roughly equal length. It begins with a chapter on the classical special functions. There are several reasons for introducing these from the beginning. One is that these functions arise frequently in many branches of mathematics, pure and applied, and a mathematician should be able, at a minimum, to recognize them and have some idea what is known about them. A second reason is that special functions are a convenient example, often the motivating example, for the general theory.

In addition to the Gamma and Beta functions and what are often called the functions of mathematical physics, [1] I have included sections on the zeta function and its implications for the distribution of primes. All of this material is presented from a "real variable" point of view. This may seem odd in a book on complex analysis, but it is not without reason. Real variable techniques, particularly those depending on convexity and Hölder's inequality, are often useful and worth knowing. They have their limits though, and we soon reach the point where further progress requires the techniques of complex analysis.

Complex analysis makes its entrance in the second chapter. This necessarily starts out rather drearily, as the various equivalent definitions of analytic and meromorphic functions are given and their elementary properties proved. In these sections I have willingly sacrificed elegance for speed. Once they are out of the way the serious work can begin.

One of the most powerful and, for the novice, most mysterious techniques of complex analysis is Cauchy's calculus of residues. This is in some sense the point of analytic and meromorphic functions. The chapter begins with the easiest examples, rational functions integrated along the unit circle. It concludes with some rather intricate calculations. I hope that these latter give some indication of the enormous power of the calculus of residues.

The third part of the book covers elliptic and modular functions. I decided to treat these in much more detail, and from a different point of view, than is usual in introductory books. I have largely followed Borchardt's notes [4] from Jacobi's Berlin lectures, still in many respects the best introductory treatment available, but with much more extensive use of the calculus of residues.

---

[1] The hypergeometric and confluent hypergeometric functions. These include as special cases the functions associated with the names of Bessel, Legendre, Laguerre, Chebyshev, and many others. In fact they occur as often outside of physics as in it. Most of the functions of statistics are of this type, as are many that appear in analytic number theory and representation theory. Conversely, many functions which are important in physics are conventionally omitted from the list of "functions of mathematical physics." For example, the zeta function and various theta functions play an important role in statistical mechanics.

The deepest applications of this material are to the theory of numbers. As an example I give Eisenstein's proof of biquadratic reciprocity. A proper treatment of complex multiplication would require some number theory, but I have sought at least to provide the necessary analytic background.

The preceding paragraphs describe what topics are covered in this book. Many more are omitted. This is, after all, an introduction. The following paragraphs provide an admittedly incomplete list of the topics omitted, my reasons for omitting them, and some places where they can be found.

A regrettable omission is the subject of conformal mapping. Most students will not miss it much; its practical importance has been diminishing for a century now. Those who do need it are better advised to look in a specialist book like Nehari [54] than a general introduction to complex analysis. Along with conformal mapping I have also excised the Riemann mapping theorem. Its main purpose is, in any case, psychological; though it provides reassurance when one is looking for a conformal map, it does not help find the map. All introductory books copy Koebe's proof, so it scarcely matters where one looks.

Runge's theorem is also missing. Like the Riemann mapping theorem it is primarily used for psychological reassurance. A fast, but non-constructive, proof is given in the first chapter of Hörmander [37]. A slow proof is given in Markushevich [53]. This latter is "constructive" by the standards of mathematical logic, but anyone interested in the practical aspects of polynomial and rational approximation is well advised to look elsewhere, and to ignore Runge's theorem entirely. Szegö [65] and Perron [57] are the standard references for approximation by polynomials and rational functions respectively. Stieltjes' article [63] is also a good place to look for the latter.

The Laplace transform appears only in examples and exercises. No attempt is made to present the general theory. The general theory divides fairly neatly into "real" and "complex" parts. The dominant technique in the former is integration by parts. The dominant technique in the latter is deforming the contour of integration. These are precisely the techniques which characterize chapters 1 and 2 of this book, so these should be good preparation for anyone who wishes to proceed further in this direction. A good specialist text on the Laplace transform is Doetsch [19]. An excellent condensed treatment may be found in Henrici [35]. Similar remarks apply to the Fourier transform. A standard reference is Titchmarsh [68].

Harmonic and subharmonic functions do not put in an appearance. Their presence will be missed mainly by those planning to continue with several complex variables and elliptic partial differential equations. These readers may wish to consult the first chapter of Hörmander [37].

There is a beautiful theory of entire functions relating their growth at infinity, the size of the coefficients in their Taylor expansions, and the distribution of their zeroes. It has had some spectacular successes in producing zeroes of functions "out of thin air," but most students will never need it. It did not seem justifiable to include it on purely aesthetic grounds. Those students who do need it may wish to look in Cartwright [6] or Levin [51].

Though the treatment of the classical special functions given here is much more detailed than is usual in an introductory text, no effort has been made to be encyclopedic. The intent is to illustrate the available techniques, not to write down every identity which can be proved by those techniques. The best general reference is the Bateman manuscript project's giant [22]. Even this is far from complete, though. For particular functions one should refer to specialized texts. Buchholz [5] is the standard reference for Whittaker functions. Watson [71] is the standard for Bessel functions. The hypergeometric functions one meets in practice are almost invariably Legendre functions. For these see Hobson [36].

Acknowledgments. I would like to thank Bob Gunning, Joe Kohn, Bert Kostant, Kamal Makdisi, and Dan Stroock for the encouragement while I was writing this book. I would also like to thank Ann Kostant and the staff at Birkhäuser for their help in preparing it for publication. Finally, I would like to thank my wife Milada and my son Anthony for their patience throughout the process.

Chapter 1
Special Functions

This chapter introduces the classical special functions. It is certainly not intended to give an exhaustive treatment of them; rather we are interested in these functions primarily as examples of technique. The problem of determining the behavior of a definite integral, or of the solution to a differential equation, as a function of various parameters is one that arises frequently. If one is lucky, one finds that someone else has encountered the same function and that a reasonably complete theory exists. More often one is left to one's own devices. For this reason it is much more important to understand the available techniques than to have memorized a list of formulae.

The techniques in this part are essentially "real variable" methods. All parameters, variables, functions, etc. are real-valued unless otherwise stated. Perhaps this seems bizarre in a complex analysis text, but there are several reasons for it. The functions in question are most interesting for these values. Some of their most interesting properties, those related to convexity, have no simple analogues in the complex theory. These properties often influence their complex behavior as well. This interaction between real and complex analysis goes in both directions. Pedagogically, the advantage is that one can can begin at once interesting examples since the requisite theoretical background is simply the differential and integral calculus.

A few words about notation may be helpful. In each section we consider a different definite integral. Rather than wasting the scarce resources of the alphabet, we call each of them I. Thus the meaning of the letter I is constant only within sections. In each case the integral in question, or some multiple of it, has a conventional name, which we introduce after proving its principal properties. In this field consistency counts for more than elegance, and we use the most widespread notation, even where another choice would have been better. The reader should regard the words "parameter" and "variable" as synonyms, although there is a pattern to their use.

J. Stalker, *Complex Analysis: Fundamentals of the Classical Theory of Functions*,
Modern Birkhäuser Classics, DOI 10.1007/978-0-8176-4919-7_1,
© Birkhäuser Boston, a part of Springer Science + Business Media, LLC 2009

As mentioned above, the necessary background for this chapter is the differential and integral calculus. The existence and uniqueness theorem for differential equations will be mentioned but not used. Since few calculus classes give proper attention to the hypotheses under which formal operations on integrals, especially improper integrals, are justified, and since most real analysis texts include much that is irrelevant to our present purposes, we provide a brief review in the appendix. A few of the results, Hölder's inequality for example, may be new to some readers and for these complete proofs are given. Readers familiar with this material will, and should, ignore the appendix.

1.1   The Gamma Function

Let the function $I(z, s)$ be defined by the following integral.[1]

$$I(z, s) = \int_0^\infty t^{z-1} e^{-st} \, dt. \tag{1}$$

The proper response to any sentence beginning with either "Let" or "Suppose" is "Why?" Sometimes one must wait many pages for an answer. Often the only answer which can be given immediately is "Trust me, eventually you will see that this object is genuinely useful." The full importance of the function $I(z, s)$ will become clear only gradually, but some justification can be given immediately.

We will see below that $I(n+1, 1)$ is just $n!$. Upper and lower bounds on $n!$ are important in estimating probabilities. A simple example is given at the end of this section. More subtly, such bounds can be used to estimate the number of primes below some given limit. This will be illustrated by Theorem 2 in Section 1.2. It is generally easier to estimate a function defined for all positive reals than one defined only for positive integers. The proof of the estimates (1.3) below uses the differential and integral calculus in an essential way. These techniques would not be available had we not extended the definition of the factorial in some way. Of course there are many different ways to extend the factorial. Theorem 1 below is intended to suggest that (1) is the right way to extend the factorial.

Another motivation for the definition (1) comes from the theory of Laplace transforms. If $f(t)$ is a function defined for t positive and real, then its *Laplace transform* $F(s)$ is defined by

$$F(s) = \int_0^\infty f(t) e^{-st} \, dt \tag{2}$$

---

[1]All integrals here and throughout the book may be interpreted either as Riemann integrals or as Lebesgue integrals, according to the reader's preference. We can avoid specifying the nature of the integral since we need never integrate a function not already known to be continuous, and all definitions of the integral agree for continuous functions. See the appendix for more detail.

for all s for which the integral is convergent. Our $I(z, s)$ then is the Laplace transform of the function t^{z-1}.

The integral (1) converges for $z > 0$ and $s > 0$ and diverges for $z \leq 0$ or $s \leq 0$. Indeed if $0 < s_1 \leq s \leq s_2$ and $0 < z_1 \leq z \leq z_2$, then

$$|t^{z-1}e^{-st}| \leq t^{z_1-1} \tag{3}$$

if $0 < t \leq 1$ and

$$|t^{z-1}e^{-st}| \leq \left(\frac{2z_2 - 2}{s_1}\right)^{z_2-1} e^{-s_1 t/2} \tag{4}$$

if $1 \leq t$. For the latter we must use the inequality (23) of the appendix. Since

$$\int_0^1 t^{z_1-1}\, dt = z_1^{-1} \tag{5}$$

and

$$\int_1^\infty e^{-s_1 t/2} = 2s_1^{-1}, \tag{6}$$

we conclude that the integral converges uniformly for $0 < s_1 \leq s \leq s_2$ and $0 < z_1 \leq z \leq z_2$ by the comparison test. It follows from Theorem R of the appendix that I is continuous in both its arguments. In general, verifying the convergence of improper integrals is a straightforward but rather tedious task, and we will give the calculation in detail only if there is some subtlety involved.

The most important property of the function $I(z, s)$ is the functional equation,

$$sI(z+1, s) = zI(z, s), \tag{7}$$

which we prove by integration by parts,

$$
\begin{aligned}
sI(z+1, s) &= \int_0^\infty t^z s e^{-st}\, dt \\
&= -\int_0^\infty t^z \frac{d}{dt} e^{-st}\, dt \\
&= \int_0^\infty \left(\frac{d}{dt} t^z\right) e^{-st}\, dt \\
&= \int_0^\infty z t^{z-1} e^{-st}\, dt \\
&= zI(z, s).
\end{aligned}
\tag{8}
$$

This formula is very useful for studying the dependence of $I(z, s)$ on the parameter z. The dependence on the variable s is much simpler. We can prove the formula

$$I(z, s) = I(z, 1)s^{-z} \tag{9}$$

by making the change of variables $u = st$ in the integral (1),

$$
\begin{aligned}
I(z, s) &= \int_0^\infty t^z e^{-st} \frac{dt}{t} \\
&= \int_0^\infty (\frac{u}{s})^z e^{-u} \frac{du}{u} \\
&= s^{-z} \int_0^\infty u^z e^{-u} \frac{du}{u} \\
&= s^{-z} I(z, 1).
\end{aligned}
\tag{10}
$$

From (9) it is easy to prove Euler's differential equation

$$(s\frac{d}{ds} + z)I(z, s) = 0. \tag{11}$$

When we consider the Whittaker and hypergeometric integrals in later sections, this change of variables will not be available, so it is worthwhile to note that there is an alternate approach to (9). We can differentiate the integral representation (1) under the integral sign to obtain

$$\frac{d}{ds}I(z, s) = -I(z + 1, s). \tag{12}$$

The differentiation under the integral sign is justified by Theorem S of the appendix The differential equation (11) follows from this and the functional equation (7). The existence and uniqueness theorem for ordinary differential equations then implies (9), though one can also give a more elementary proof.

Some special values of the integral (1) are easier to compute than others. In particular

$$I(1, s) = \frac{1}{s}. \tag{13}$$

is an easy exercise in the integral calculus. There is also a trick for $z = \frac{1}{2}$, but we will defer this for a few pages.

The next property of $I(z, s)$ lies somewhat deeper. The following inequality is known as log convexity:

$$I(\alpha z + \alpha' z', \alpha s + \alpha' s') \le I(z, s)^\alpha I(z', s')^{\alpha'} \tag{14}$$

for $\alpha, \alpha', z, z', s, s' > 0$ and $\alpha + \alpha' = 1$. That is an immediate consequence of Hölder's inequality. In (73) of the appendix we take

$$\begin{aligned} f(t) &= t^{z-1}e^{-st} \\ g(t) &= t^{z'-1}e^{-s't}. \end{aligned}$$

In some sense we now know everything there is to know about $I(z, s)$. The properties (7), (9), (13), and (14) uniquely determine it. That is the content of the following theorem. Since the dependence on s is so trivial, it suffices to consider only $I(z, 1)$.

THEOREM 1 (UNIQUENESS OF GAMMA)
There is one and only one function $\Gamma(z)$ *with the following three properties:*

  1. Functional Equation:

$$\Gamma(z + 1) = z\Gamma(z). \tag{15}$$

  2. Special Value:

$$\Gamma(1) = 1. \tag{16}$$

  3. Log Convexity:

$$\Gamma(\alpha z + \alpha' z') \le \Gamma(z)^{\alpha}\Gamma(z')^{\alpha'} \tag{17}$$

  for $\alpha + \alpha' = 1$.

This theorem is intended mostly for psychological reassurance. It tells us that the Gamma function is in some sense the right extension of the factorial. It does have a practical use however. There are many representations of the Gamma function. We will soon see several of them. To prove that they are all equivalent, it suffices to check the hypotheses (15), (16), and (17) for each of them.

Proof. We have already established the existence of such a Γ since we have seen that

$$\Gamma(z) = I(z, 1) \tag{18}$$

has all the desired properties. We therefore proceed to uniqueness. First we need the following improved version of the functional equation:

$$\Gamma(z + n) = (z)_n\Gamma(z), \tag{19}$$

where

$$(z)_n = \prod_{j=0}^{n-1}(z+j), \tag{20}$$

a notation, known as the *Pochhammer symbol*, which is very useful when dealing with the Gamma function and its cousins. It is occasionally useful to extend the Pochhammer symbol to negative values of the subscript. This is done by setting

$$(z)_{-n} = \frac{1}{(z-n)_n}. \tag{21}$$

Equation (19) is vacuously true for $n = 0$, and the general case follows by induction from (15). The special value (16) and the generalized functional equation (19) combine to give the promised relation

$$\Gamma(n+1) = n! \tag{22}$$

between the Gamma function and the factorial.

The following limit formula constitutes the heart of the proof. For any b, c,

$$\lim_{u\to\infty} \frac{u^b\Gamma(u+c)}{u^c\Gamma(u+b)} = 1. \tag{23}$$

The limit formula is obvious for $b = c$ and is unchanged on switching the roles of b and c, so it suffices to consider the case $b < c$. It is clear that we can find a, d such that $a < b < c < d$ and the two differences $c - a$ and $d - b$ are integers.

We now apply log convexity twice. The first time we take

$$z = a + u \qquad\qquad \alpha = \frac{c-b}{c-a} \tag{24}$$

$$z' = c + u \qquad\qquad \alpha' = \frac{b-a}{c-a}$$

and we find

$$\Gamma(u+b) \leq \Gamma(u+a)^{\frac{c-b}{c-a}}\Gamma(u+c)^{\frac{b-a}{c-a}} \tag{25}$$

or, using the improved functional equation (19),

$$\Gamma(u+b) \leq (u+a)^{\frac{b-c}{c-a}}\Gamma(u+c) \tag{26}$$

We rewrite the preceding inequality as

$$\left(1+\frac{a}{u}\right)^{\frac{c-b}{c-a}}\left(1+\frac{a+1}{u}\right)^{\frac{c-b}{c-a}}\cdots\left(1+\frac{c-1}{u}\right)^{\frac{c-b}{c-a}} \leq \frac{u^b\Gamma(u+c)}{u^c\Gamma(u+b)}, \tag{27}$$

and note that the quantity on the left tends to 1 as u tends to infinity. Next we take

$$z = b + u \qquad\qquad \alpha = \frac{d-c}{d-b} \qquad\qquad (28)$$

$$z' = d + u \qquad\qquad \alpha' = \frac{c-b}{d-b}$$

and find

$$\Gamma(u+c) \le \Gamma(u+b)^{\frac{d-c}{d-b}}\Gamma(u+d)^{\frac{c-b}{d-b}} \qquad (29)$$

or, again using (19),

$$\Gamma(u+c) \le (u+b)^{\frac{c-b}{d-b}}\Gamma(u+b), \qquad (30)$$

which we rewrite as

$$\frac{u^b \Gamma(u+c)}{u^c \Gamma(u+b)} \le (1+\frac{b}{u})^{\frac{c-b}{d-b}}(1+\frac{b+1}{u})^{\frac{c-b}{d-b}}\cdots(1+\frac{d-1}{u})^{\frac{c-b}{d-b}}, \qquad (31)$$

and note that the quantity on the right tends to 1 as u tends to infinity. Since

$$\frac{u^b \Gamma(u+c)}{u^c \Gamma(u+b)}$$

is squeezed between two bounds, both of which tend to 1 as u tends to infinity, the limit formula (23) is proved.

As a first application we take $b = z+1$ and $c = 1$ and restrict u to integer values n. We find

$$\lim_{n\to\infty} \frac{n^z \Gamma(n+1)}{\Gamma(z+n+1)} = 1. \qquad (32)$$

Using the functional equation (19) and the special value (22) we can rewrite this as

$$\lim_{n\to\infty} \frac{n^z n!}{(z)_{n+1}\Gamma(z)} = 1 \qquad (33)$$

or

$$\Gamma(z) = \lim_{n\to\infty} \frac{n^z n!}{z(z+1)\cdots(z+n)} \qquad (34)$$

The representation (34) appears in Gauss' paper [31] on the hypergeometric function.[2] It determines the value of $\Gamma(z)$ uniquely, and we have used only the hypotheses (15), (16), and (17) of the theorem in its proof, so the theorem is proved.

This theorem would be a mere curiosity but for three observations. First, the integral representation (1) with which we started is not the same as the limit representation (34) which we found above, so the proof of the theorem implies the nontrivial identity

$$\int_0^\infty t^{z-1} e^{-t}\, dt = \lim_{n\to\infty} \frac{n^z n!}{z(z+1)\cdots(z+n)}, \tag{35}$$

valid for $z > 0$. Second, although the integral representation ceases to be meaningful as soon as $z \leq 0$, the limit representation is well-defined and finite, provided that z is not a nonpositive integer. For this reason it is advisable to take the limit representation (34) as the definition of the Gamma function, and consider formula (18) as a theorem, valid only under the hypothesis $z > 0$. Extending functions beyond the regions on which they are initially defined is something of a recurring theme in complex analysis. Third, the limit formula (23) is of independent interest. We will use it below in a simple probability problem. Later it will play a role in the distribution of primes.

What are the odds of flipping a fair coin $2n$ times and getting heads exactly as often as tails? The answer is clearly

$$2^{-2n} \frac{(2n)!}{(n!)^2},$$

but this formula does not make the behavior for large n sufficiently clear.

We want to determine the behavior of this probability as n becomes large. Using the Pochhammer symbol we can write

$$n! = \prod_{j=0}^{n-1} (j+1) = (1)_n. \tag{36}$$

We can also write

$$(2n)! = \prod_{j=0}^{2n-1} (j+1) = (1)_{2n}, \tag{37}$$

but this is less useful. Instead we break up the product into two pieces. The first will contain all the odd j's, those of the form $j = 2k + 1$ with $0 \leq k \leq n - 1$.

---

[2]The formula was certainly known earlier. It appears in a letter of Euler to Goldbach dated 1729. Dirichlet claims in [16] that the representation (34) was first published by Laplace. No doubt he is correct, but the reference is too vague to trace. The name Gauss limit representation has therefore stuck.

The second will contain all the even j's, those of the form $j = 2k + 2$ with $0 \leq k \leq n - 1$. In this way we find

$$(2n)! = \prod_{k=0}^{n-1}(2k + 1) \prod_{k=0}^{n-1}(2k + 2). \tag{38}$$

Now we extract a 2 from each factor,

$$(2n)! = 2^{2n} \prod_{k=0}^{n-1}(k + \frac{1}{2}) \prod_{k=0}^{n-1}(k + 1) = 2^{2n}(\frac{1}{2})_n (1)_n. \tag{39}$$

From (36) and (39) we see that we can rewrite the probability in the form

$$2^{-2n}\frac{(2n)!}{(n!)^2} = \frac{(\frac{1}{2})_n}{(1)_n}. \tag{40}$$

Now we are in a position to use the limit formula (23). If we take $b = \frac{1}{2}$ and $c = 1$, we find

$$\lim_{n \to \infty} \frac{\Gamma(n + 1)}{n^{1/2}\Gamma(n + \frac{1}{2})} = 1 \tag{41}$$

or, using (19)

$$\lim_{n \to \infty} \frac{(1)_n \Gamma(1)}{n^{1/2}(\frac{1}{2})_n \Gamma(\frac{1}{2})} = 1. \tag{42}$$

In other words the probability of getting n heads and n tails is approximately

$$\frac{\Gamma(1)}{\Gamma(\frac{1}{2})} n^{-1/2}.$$

We already know that $\Gamma(1)$ is just 1. The following trick allows us to evaluate $\Gamma(\frac{1}{2})$ directly from the integral representation (18).

$$\begin{aligned}
\Gamma(\tfrac{1}{2})^2 &= \int_0^\infty u^{-1/2}e^{-u}\, du \int_0^\infty v^{-1/2}e^{-v}\, dv \tag{43}\\
&= 4\int_0^\infty e^{-x^2}\, dx \int_0^\infty e^{-y^2}\, dy \\
&= 4\int_0^\infty \int_0^\infty e^{-x^2-y^2}\, dx\, dy \\
&= 4\int_0^{\pi/2} \int_0^\infty re^{-r^2}\, dr\, d\theta \\
&= 2\int_0^{\pi/2} \int_0^\infty e^{-t}\, dt\, d\theta \\
&= \pi.
\end{aligned}$$

From the integral representation we see that $\Gamma(z) > 0$ for $z > 0$, so

$$\Gamma(\frac{1}{2}) = \sqrt{\pi}. \tag{44}$$

Equation (42) says that

$$\frac{(\frac{1}{2})_n}{(1)_n} = \frac{1}{\sqrt{\pi n}} + o(n^{-1/2}). \tag{45}$$

The estimates

$$\frac{1}{\sqrt{\pi(n+\frac{1}{2})}} \leq \frac{(\frac{1}{2})_n}{(1)_n} \leq \frac{1}{\sqrt{\pi n}}, \tag{46}$$

which are left as an exercise, give sharper bounds. They can be used to compute the probability to within an error proportional to $n^{-3/2}$.

Exercises

1. Prove the estimates (46). *Hint:* look closely at the proof of the limit formula (23).

2. Using the same technique that produced (44), prove the following formula of Euler:

$$2 \int_0^{\pi/2} \sin^{2p-1}\theta \cos^{2q-1}\theta \, d\theta = \frac{\Gamma(p)\Gamma(q)}{\Gamma(p+q)},$$

   for $p, q > 0$.

3. Prove the Legendre duplication formula

$$\Gamma(z)\Gamma(z+\frac{1}{2}) = 2^{1-2z}\pi^{1/2}\Gamma(2z).$$

   You may find Theorem 1 useful.

4. Suppose $F(s)$ and $G(s)$ are the Laplace transforms of $f(t)$ and $g(t)$ respectively, where

$$g(t) = \frac{d}{dt}f(t).$$

   Show that

$$G(s) = sF(s) - f(0)$$

   What is the relation between this formula and the functional equation (7)?

5. The most important fact about the Gamma function missing from our discussion is the formula

$$\Gamma(z)\Gamma(1-z) = \pi \csc \pi z. \tag{47}$$

Try to prove this formula. Don't be surprised if you can't. Collect as much information as you can about the function on the left and show that this information is consistent with the formula.

6. Prove that the integral $I(z, s)$ is infinitely differentiable in both its arguments and express

$$\frac{d^k}{dz^k} I(z, s)$$

as an integral.

7. Show that $\Gamma(z)$ is never zero and that $1/\Gamma$ extends to an infinitely differentiable function on the whole real line, even at the nonpositive integers, where Gamma is undefined.

1.2 The Distribution of Primes I

It is perhaps surprising that we already have enough analysis at our disposal to prove nontrivial results in number theory. Consider the functions

$$\pi(x) = \sum_{p \leq x} 1, \tag{1}$$

$$\operatorname{li} x = \int_0^x \frac{du}{\log u}.$$

Here and throughout this section p ranges over primes. Clearly $\pi(x)$ is just the number of primes less than or equal to x. If $x > 1$, then the integrand has a discontinuity at $t = 1$. The definition above should then be understood to mean

$$\operatorname{li} x = \lim_{\epsilon \to 0^+} \left(\int_0^{1-\epsilon} \frac{du}{\log u} + \int_{1+\epsilon}^x \frac{du}{\log u} \right)$$

in this case. This function first appears in Euler's [27]. It is a special case of the Whittaker function, which we will meet in Section 1.5.

The main results on the distribution of primes are that $\pi(x) - \operatorname{li} x$ is never very large and that it is often very small. The first results in both directions are due to Chebyshev, and it is these which we will prove in this section and in Section 1.9. To prove that $\pi(x) - \operatorname{li} x$ is never very large, we need only the Gamma function.

To prove that $\pi(x) - \mathrm{li}\, x$ is often very small, we will need the zeta function as well.

Chebyshev proved the following theorem with explicit constants c and C in his paper [10].

THEOREM 2 (CHEBYSHEV'S SECOND THEOREM)
The estimates

$$c\,\mathrm{li}\,x \le \pi(x) \le C\,\mathrm{li}\,x \tag{2}$$

hold for all sufficiently large x.

De la Vallée Poussin in [13] and Hadamard in [33] independently improved this to

$$\lim_{x \to \infty} \frac{\pi(x)}{\mathrm{li}\,x} = 1. \tag{3}$$

In other words we can take $c = 1 - \epsilon$ and $C = 1 + \epsilon$ for any positive ϵ. Their proof used the behavior of the zeta function for complex s. Erdős and Selberg in [23] and [61] gave "real variable" proofs without the zeta function. Conjecturally

$$\pi(x) = \mathrm{li}\,x + O(x^{1/2}\log x),$$

but this seems to require the Riemann hypothesis. Here we prove only Chebyshev's result, with a weaker pair of constants c and C, but a shorter proof.

Proof.   The main idea is quite simple. We can easily determine which primes divide $\binom{2n}{n}$ and how often. Since we know the behavior of $\binom{2n}{n}$ for large n, we can determine something about the distribution of primes.

The first step is to show that $\mathrm{li}\,x$ may be replaced by $x \log^{-1} x$ in Theorem 2. Exercise 3 will examine the behavior of $\mathrm{li}\,x$ for large x in some detail. Here we need only weak estimates, so we can afford to be a little lazy. We begin by integrating by parts. Let $a > 1$ be any constant, fixed for the remainder of the discussion.

$$
\begin{aligned}
\mathrm{li}\,x &= \mathrm{li}\,a + \int_a^x \frac{du}{\log u} \\
&= \mathrm{li}\,a + \frac{x}{\log x} - \frac{a}{\log a} - \int_a^x \frac{du}{\log^2 u} \\
&= \mathrm{li}\,a + \frac{x}{\log x} - \frac{a}{\log a} - \int_a^y \frac{du}{\log^2 u} - \int_y^x \frac{du}{\log^2 u}
\end{aligned}
\tag{4}
$$

The first and third terms on the right are constants. If we take $y = x \log^{-2} x$ then the fourth and fifth terms are both $O(x \log^{-2} x)$ and hence $o(x \log^{-1} x)$. It follows that

$$\lim_{x \to \infty} \frac{\text{li } x}{x \log^{-1} x} = 1. \tag{5}$$

The next step is to determine the prime factorization of $\binom{2n}{n}$. p divides $m!$ precisely

$$\sum_{k=1}^{\infty} [\frac{m}{p^k}]$$

times where $[z]$ represents the greatest integer not exceeding z. This formula simply expresses the fact that $[\frac{m}{p}]$ of the numbers less than or equal to m are divisible by p, $[\frac{m}{p^2}]$ are divisible by p^2, $[\frac{m}{p^3}]$ by p^3, etc. We will need the following simple property of the greatest integer function

$$[z] - 2[\frac{z}{2}] = \sum_{1 \le l \le z} (-1)^{l+1}. \tag{6}$$

Indeed, it is clear that both sides are either 0 or 1 according to whether $[z]$ is even or odd. The number of times p divides $\binom{2n}{n}$ is given by

$$\sum_{k=1}^{\infty} \left([\frac{2n}{p^k}] - 2[\frac{n}{p^k}] \right), \tag{7}$$

or, using the identity (6),

$$\sum_{k,l} (-1)^{l+1}. \tag{8}$$

The sum here is over all positive integers k and l such that $lp^k \le 2n$. Writing $\binom{2n}{n}$ as a product of primes, we find

$$\log \binom{2n}{n} = \sum_{p,k,l} (-1)^{l+1} \log p. \tag{9}$$

This is a finite sum, so we may rearrange it at will. We choose to sum over k, then over p, and finally over l. Summing over k gives

$$\log \binom{2n}{n} = \sum_{p,l} (-1)^{l+1} [\frac{\log(2n/l)}{\log p}] \log p. \tag{10}$$

We can take this sum to be over all primes p and all positive integers l, though only finitely many terms will be nonzero. Summation over p gives

$$\log \binom{2n}{n} = \sum_l (-1)^{l+1} \psi(2n/l), \tag{11}$$

where the function ψ is defined by

$$\psi(x) = \sum_{p \leq x} [\frac{\log x}{\log p}] \log p. \tag{12}$$

The sum over l gives

$$\log \binom{2n}{n} = \omega(2n), \tag{13}$$

where the function ω is defined by

$$\omega(x) = \sum_{l=1}^{\infty} (-1)^{l+1} \psi(x/l). \tag{14}$$

For future reference we note that the argument given above for (13) gives slightly more information. We can break the argument into two parts. First we showed that

$$\log \binom{2n}{n} = \sum_{p,k} \log p \tag{15}$$

where the sum is over primes p and integers k such that $[\frac{2n}{p^k}]$ is odd. Then we showed that

$$\sum_{p,k} \log p = \omega(x) \tag{16}$$

where the sum is over primes p and integers k such that $[\frac{x}{p^k}]$ is odd. This is meaningful even when x is not an even integer.

The next step is to prove that

$$\lim_{x \to \infty} \frac{\omega(x)}{x} = \log 2. \tag{17}$$

When $x = 2n$ we can use the estimate (46) from Section 1.1. This gives

$$L(x) \leq \omega(x) \leq U(x) \tag{18}$$

when x is an even integer. The upper and lower bounds are given by

$$L(x) = x \log 2 - \frac{1}{2} \log(x + \frac{1}{4}) - \frac{1}{2} \log \frac{\pi}{2} \qquad (19)$$

$$U(x) = x \log 2 - \frac{1}{2} \log x - \frac{1}{2} \log \frac{\pi}{2}.$$

Note that the upper and lower bounds differ by a quantity which is $O(x^{-1})$ for x large. Unfortunately we cannot maintain this level of precision in the remainder of the proof.

Rather annoyingly, the function ω is neither an increasing nor a decreasing function. One can see, though, from equation (16) that it changes fairly slowly. In particular it can change only at integers. At an integer n it can change by no more than $\log n$. The functions L and U change even more slowly. They can change only by an amount which is $O(1)$ when we increase x by 1. Of course every x is within a distance 1 of the nearest even integer. From the estimates (18), which hold when x is an even integer, we can therefore conclude that

$$L(x) - O(\log x) \le \omega(x) \le U(x) + O(\log x), \qquad (20)$$

with no restriction on x. We note that

$$L(x) = x \log 2 + O(\log x) \qquad (21)$$

and

$$U(x) = x \log 2 + O(\log x). \qquad (22)$$

It follows that

$$\omega(x) = x \log 2 + O(\log x) \qquad (23)$$

and (17) is proved.

The next step is to convert our estimates on ω into estimates on ψ. This is the step where we lose most of our information. The sum (14), used to define ω, is a sum of decreasing terms with alternating signs. We can rearrange it in either of two ways. We can write

$$\omega(x) = \psi(x) - \sum_{k=1}^{\infty} \left(\psi(\frac{x}{2k}) - \psi(\frac{x}{2k+1}) \right) \qquad (24)$$

or

$$\omega(x) = \psi(x) - \psi(\frac{x}{2}) + \sum_{k=1}^{\infty} \left(\psi(\frac{x}{2k+1}) - \psi(\frac{x}{2k+2}) \right). \qquad (25)$$

In either case the sum contains only nonnegative terms and must therefore be nonnegative. This shows that

$$\psi(x) - \psi(\frac{x}{2}) \le \omega(x) \le \psi(x). \tag{26}$$

Summation by parts, equation (68) of the appendix, shows that

$$\sum_{l=0}^{\infty} (\psi(2^{-l}x) - \psi(2^{-l-1}x)) = \psi(x), \tag{27}$$

and hence

$$\omega(x) \le \psi(x) \le \sum_{l=0}^{\infty} \omega(2^{-l}x). \tag{28}$$

From (17) we see that, for any positive ϵ,

$$x(1 - \epsilon) \log 2 \le \omega(x) \le x(1 + \epsilon) \log 2 \tag{29}$$

provided x is greater than some bound $X(\epsilon)$. We can break the sum in (28) into two pieces, those for which $2^{-l}x > X(\epsilon)$ and those for which $2^{-l}x \le X(\epsilon)$. Of the latter, at most

$$[\frac{\log X(\epsilon)}{\log 2}]$$

can be nonzero. None of them can be greater than

$$\max_{y \le X(\epsilon)} \omega(y).$$

Thus we see that the second piece is bounded independently of x. The first piece is bounded by

$$(1 + \epsilon) \sum_{l=0}^{\infty} 2^{-l}x \log 2$$

or

$$x(1 + \epsilon)2 \log 2.$$

It follows that

$$\psi(x) \le x(1 + \epsilon)2 \log 2 + O(1). \tag{30}$$

A similar but much easier argument shows that

$$x(1 - \epsilon) \log 2 - O(1) \le \psi(x). \tag{31}$$

The $O(1)$ terms can be absorbed if we are willing to increase $X(\epsilon)$. We conclude that

$$x(1 - \epsilon) \log 2 \le \psi(x) \le x(1 + \epsilon)2 \log 2 \tag{32}$$

for all sufficiently large x.

The final step is to estimate $\pi(x)$. $\psi(x)$ is defined as a sum of $\pi(x)$ terms, each bounded by $\log x$, so

$$\psi(x) \le \pi(x) \log x. \tag{33}$$

On the other hand, if we omit terms with $p < y$, then we are left with at least $\pi(x) - y$ terms of size at least $\log y$, so

$$(\pi(x) - y) \log y \le \psi(x) \tag{34}$$

or

$$\pi(x) \le \frac{\psi(x)}{\log y} + y. \tag{35}$$

We take $y = \frac{x}{\log^2 x}$ and find

$$\frac{\psi(x)}{\log x} \le \pi(x) \le \frac{\psi(x)}{\log x}(1 - 2\frac{\log \log x}{\log x})^{-1} + \frac{x}{\log^2 x}. \tag{36}$$

Using our estimates (32) for ψ, we can write this as

$$(1 - \epsilon) \log 2 \frac{x}{\log x} \le \pi(x) \le (1 + \epsilon)2 \log 2 \frac{x}{\log x}(1 - 2\frac{\log \log x}{\log x})^{-1} + \frac{x}{\log^2 x}. \tag{37}$$

This holds for any sufficiently large x.

From (37) we see that

$$(1 - \epsilon) \log 2 \le \liminf_{x \to \infty} \frac{\pi(x) \log x}{x} \le \limsup_{x \to \infty} \frac{\pi(x) \log x}{x} \le (1 + \epsilon)2 \log 2. \tag{38}$$

This must hold for any positive ϵ so

$$\log 2 \le \liminf_{x \to \infty} \frac{\pi(x) \log x}{x} \le \limsup_{x \to \infty} \frac{\pi(x) \log x}{x} \le 2 \log 2. \tag{39}$$

Using (5), we see that this is equivalent to

$$\log 2 \le \liminf_{x \to \infty} \frac{\pi(x)}{\operatorname{li} x} \le \limsup_{x \to \infty} \frac{\pi(x)}{\operatorname{li} x} \le 2 \log 2. \tag{40}$$

Thus (2) is true for any $c < \log 2$ and $2 \log 2 < C$.

Exercises

1. Where does ω decrease?

2. One of Chebyshev's main goals was to disprove a "theorem" stated without proof by Legendre in [50]. For this purpose he needed one more term in the expansion of $\operatorname{li} x$ than we have given above. Show that

$$\operatorname{li} x = \frac{x}{\log x} + \frac{x}{\log^2 x} + O\left(\frac{x}{\log^3 x}\right).$$

   You will need to integrate by parts once more before splitting the integral.

3. The idea of the preceding exercise can be carried further. Show that

$$\operatorname{li} x = \sum_{j=0}^{m-1} a_j \frac{x}{\log^j x} + O\left(\frac{x}{\log^m x}\right).$$

   What are the constants a_j? Could these be the partial sums in a convergent series expansion

$$\operatorname{li} x = \sum_{j=0}^{\infty} a_j \frac{x}{\log^j x}?$$

4. Chebyshev's proof used

$$\frac{(30n)! \, n!}{(15n)! (10n)! (6n)!}$$

   instead of $\binom{2n}{n}$. This gives constants

$$c < \frac{1}{2} \log 2 + \frac{1}{3} \log 3 + \frac{1}{5} \log 5 - \frac{1}{30} \log 30 \approx 0.921$$

   and

$$C > \frac{6}{5}(\frac{1}{2} \log 2 + \frac{1}{3} \log 3 + \frac{1}{5} \log 5 - \frac{1}{30} \log 30) \approx 1.106$$

   in place of our

$$c < \log 2 \approx 0.692$$

   and

$$C > 2 \log 2 \approx 1.384.$$

   Reconstruct Chebyshev's proof.

1.3 Stirling's Series

Equations (18) and (34) of Section 1.1 are by no means the only representations of the Gamma function. A very interesting representation of $\Gamma(z)$ is Stirling's formula,[3]

$$\Gamma(z) = \sqrt{\frac{2\pi}{z}}(\frac{z}{e})^z e^{J(z)} \tag{1}$$

where

$$J(z) = \frac{1}{12z} - \frac{1}{360z^3} + \frac{1}{1260z^5} - \frac{1}{1680z^7} + \frac{1}{1188z^9} - \frac{691}{360360z^{11}} + \cdots . \tag{2}$$

The series (2) has many remarkable properties, but convergence is not among them. Nonetheless Stirling's representation provides a very good description of the behavior of $\Gamma(z)$ for z large. In this section we will prove the estimates

$$\sqrt{\frac{2\pi}{z}}(\frac{z}{e})^z \le \Gamma(z) \le \sqrt{\frac{2\pi}{z}}(\frac{z}{e})^z e^{\frac{1}{12z}} \tag{3}$$

for $z > 0$. The estimates

$$\sqrt{\frac{2\pi}{z}}(\frac{z}{e})^z e^{\frac{1}{12z} - \frac{1}{360z^3}} \le \Gamma(z) \le \sqrt{\frac{2\pi}{z}}(\frac{z}{e})^z e^{\frac{1}{12z}} \tag{4}$$

are left as an exercise. With a little more work one can show that

$$\sqrt{\frac{2\pi}{z}}(\frac{z}{e})^z e^{\frac{1}{12z} - \frac{1}{360z^3}} \le \Gamma(z) \le \sqrt{\frac{2\pi}{z}}(\frac{z}{e})^z e^{\frac{1}{12z} - \frac{1}{360z^3} + \frac{1}{1260z^5}} . \tag{5}$$

It is not too hard to guess what is happening, but we will stop at (3) in this section. In Section 1.7, when we have more technique at our disposal, we will return to this problem.

The product representation

$$\Gamma(z) = \frac{e^{-\gamma z}}{z} \prod_{n=1}^{\infty} (1 + \frac{z}{n})^{-1} e^{z/n} \tag{6}$$

first appears in Newman's paper [55].[4] The formula itself is of only minor importance, but we give a proof for two reasons. This representation is often given

---

[3]This representation first appeared in Stirling's book [64], although it was in some sense anticipated by De'Moivre's paper[14].

[4]Most modern books mistakenly site Weierstrass' paper [73], written eight years later, and the formula has become known as the Weierstrass product representation.

in modern texts as the definition of the Gamma function. More importantly, the proof is nearly identical to that of (3).

The idea behind the proof of (6) does not, at first sight, seem to be very promising. We start with Gauss' limit representation, equation (34) of the preceding section. We take logarithms, differentiate twice, integrate twice, and then exponentiate. We should get (34) back again, but a miracle happens and we do not.

Assume that $z > 0$. We take logarithms in (34),

$$\log \Gamma(z) = \lim_{n \to \infty} (\log n! + z \log n - \log z - \cdots - \log(z + n)). \tag{7}$$

If we formally take the limit as z tends to zero from the right we find

$$\lim_{z \to 0^+} (\log \Gamma(z) + \log z) = 0. \tag{8}$$

This, however, involves switching the order of the limits with respect to n and z and requires some effort to justify. Fortunately there is an easier way. From the functional equation (15) of the preceding section, we see that

$$\lim_{z \to 0^+} z\Gamma(z) = \lim_{z \to 1^+} \Gamma(z). \tag{9}$$

Since Γ is continuous at 1 it follows that

$$\lim_{z \to 0^+} z\Gamma(z) = \Gamma(1) = 1. \tag{10}$$

Here we have used equation (16) of Section 1.1. Taking logarithms gives exactly (8).

Differentiating (7) once,

$$\frac{d}{dz} \log \Gamma(z) = \lim_{n \to \infty} (\log n - \frac{1}{z} - \cdots - \frac{1}{z + n}). \tag{11}$$

To justify the differentiation within the limit using Theorem O of the appendix, we must check that the limit on the right hand side is locally uniformly convergent. This is fairly straightforward. What we must check is that the functions

$$f_n(z) = \log n - \sum_{j=0}^{n} (z + j)^{-1} \tag{12}$$

converge uniformly on any closed interval $[a, b]$ with $a > 0$. Since

$$\begin{aligned}
f_n(z) - f_{n-1}(z) &= (z + n)^{-1} + \log(1 - n^{-1}) \\
&= n^{-1}(1 + zn^{-1})^{-1} + \log(1 - n^{-1}).
\end{aligned} \tag{13}$$

Taylor's theorem shows that

$$|f_n(z) - f_{n-1}(z)| = O(n^{-2}) \tag{14}$$

uniformly on $[a, b]$, and the convergence of f_n follows from the comparison test.

Taking the limit as z approaches zero, we find

$$\lim_{z \to 0^+} \frac{d}{dz} \log \Gamma(z) + \frac{1}{z} = -\gamma \tag{15}$$

where γ, traditionally called Euler's constant, is defined by

$$\gamma = \lim_{n \to \infty} (1 + \frac{1}{2} + \cdots + \frac{1}{n} - \log n). \tag{16}$$

Later we will need to know the behavior as z approaches infinity. This can be deduced from the limit formula (23) of the preceding section as follows. Taking logarithms in (23) produces

$$\lim_{u \to \infty} [\log \Gamma(u + c) - \log \Gamma(u + b) - c \log u + b \log u] = 0. \tag{17}$$

The fundamental theorem of calculus allows us to rewrite this as

$$\lim_{u \to \infty} [\int_{u+b}^{u+c} \frac{d}{dz} \log \Gamma(z) \, dz - (c - b) \log u] = 0. \tag{18}$$

In some sense then the average value of $\frac{d}{dz} \log \Gamma(z)$ is approximately $\log z$ when z is large. In a few pages we will be able to make this statement much more precise.

Differentiating (11) again,

$$\frac{d^2}{dz^2} \log \Gamma(z) = \lim_{n \to \infty} \left(\frac{1}{z^2} + \cdots + \frac{1}{(z + n)^2} \right) \tag{19}$$

or, more simply,

$$\frac{d^2}{dz^2} \log \Gamma(z) = \sum_{n=0}^{\infty} (z + n)^{-2}. \tag{20}$$

Before proceeding we pause for a few quick remarks. From (20) we see that

$$\frac{d^2}{dz^2} \log \Gamma(z) > 0. \tag{21}$$

The inequality (21) seems a better candidate for the title of log convexity than equation (14) of Section 1.1. In fact, the two are equivalent. Second, the quantity on the right in (20) tends to zero as z tends to infinity. In this sense the log convexity inequality is nearly sharp for large z. This helps to explain why the peculiar argument given for the limit formula worked. It will also prove helpful when we prove (3).

We now resume the proof of the product representation. Subtracting z^{-2} from both sides of (20) and integrating from $z = 0$, we find

$$\frac{1}{z} + \frac{d}{dz} \log \Gamma(z) = -\gamma + \sum_{n=1}^{\infty} (\frac{1}{n} - \frac{1}{z+n}), \qquad (22)$$

the interchange of the sum and integral being justified by Theorem Q of the appendix. We have used (15) to determine the constant of integration. Integrating once again,

$$\log z + \log \Gamma(z) = -\gamma z + \sum_{n=1}^{\infty} (\frac{z}{n} - \log(z+n) + \log n). \qquad (23)$$

The constant of integration has been determined using (8). Exponentiating this gives the product representation (6).

We have proved the product representation only under the assumption that $z > 0$, even though both sides of the equation make sense as long as z is not a nonpositive integer. To prove (6) in general we have two choices. We can check that the right hand side satisfies the hypotheses of Theorem 1. This is one of the exercises at the end of this section. Or we can wait for the rigidity theorem of the next chapter.

The proof of (6) is more interesting than the final result. We can use the same idea to prove (3). Roughly speaking, the difference is that now we integrate from $+\infty$ rather than from 0.

Our starting point is formula (20), but this time we rewrite the sum using the identity

$$(z+j)^{-2} = (z+j)^{-1} - (z+j+1)^{-1} + (z+j)^{-2}(z+j+1)^{-1} \qquad (24)$$

We sum this using equation (68) of the appendix,

$$\sum_{j=0}^{\infty} (z+j)^{-2} = z^{-1} + \sum_{j=0}^{\infty} (z+j)^{-2}(z+j+1)^{-1}. \qquad (25)$$

Next we use the identity

$$(z+j)^{-2}(z+j+1)^{-1} = \frac{1}{2}(z+j)^{-2} - \frac{1}{2}(z+j+1)^{-2} \qquad (26)$$

$$+ \frac{1}{2}(z+j)^{-2}(z+j+1)^{-2}$$

to get

$$\sum_{j=0}^{\infty}(z+j)^{-2} = z^{-1} + \frac{1}{2}z^{-2} + \frac{1}{2}\sum_{j=0}^{\infty}(z+j)^{-2}(z+j+1)^{-2}. \qquad (27)$$

A sum of positive terms is positive so

$$\sum_{j=0}^{\infty}(z+j)^{-2} \geq z^{-1} + \frac{1}{2}z^{-2}. \qquad (28)$$

We can also apply summation by parts to (27), using the identity

$$(z+j)^{-2}(z+j+1)^{-2} = \frac{1}{3}(z+j)^{-3} - \frac{1}{3}(z+j+1)^{-3} \qquad (29)$$
$$- \frac{1}{3}(z+j)^{-3}(z+j+1)^{-3}.$$

This gives

$$\sum_{j=0}^{\infty}(z+j)^{-2} = z^{-1} + \frac{1}{2}z^{-2} + \frac{1}{6}z^{-3} - \frac{1}{6}\sum_{j=0}^{\infty}(z+j)^{-3}(z+j+1)^{-3}. \qquad (30)$$

Using, once again, the fact that a sum of positive terms is positive, we obtain the estimate

$$\sum_{j=0}^{\infty}(z+j)^{-2} \leq z^{-1} + \frac{1}{2}z^{-2} + \frac{1}{6}z^{-3} \qquad (31)$$

Combining the bounds (28) and (31) obtained above and equation (20), we find

$$z^{-1} + \frac{1}{2}z^{-2} \leq \frac{d^2}{dz^2}\log\Gamma(z) \leq z^{-1} + \frac{1}{2}z^{-2} + \frac{1}{6}z^{-3}, \qquad (32)$$

or, more succinctly,

$$0 \leq \frac{d^2}{dz^2}\log\Gamma(z) - z^{-1} - \frac{1}{2}z^{-2} \leq \frac{1}{6}z^{-3}. \qquad (33)$$

Now we begin to integrate. The upper and lower bounds in (32) agree at infinity, which strongly suggests that we should integrate from infinity. This requires some care since we are working with improper integrals. We define

$$f(z) = \frac{d}{dz}\log\Gamma(z) - \log z + \frac{1}{2}z^{-1} \qquad (34)$$

so that equation (33) becomes

$$0 \le f'(z) \le \frac{1}{6}z^{-3} \tag{35}$$

Integrating from z_1 to z_2 we find

$$0 \le f(z_2) - f(z_1) \le \frac{1}{12}(z_1^{-2} - z_2^{-2}) \le \frac{1}{12}z_1^{-2}, \tag{36}$$

so

$$c = \lim_{z \to \infty} f(z) \tag{37}$$

exists by Cauchy's criterion and

$$-\frac{1}{12}z^{-2} \le f(z) - c \le 0. \tag{38}$$

Next we define

$$g(z) = \log \Gamma(z) - (z - \frac{1}{2}) \log z + z - cz \tag{39}$$

so that equation (38) becomes

$$-\frac{1}{12}z^{-2} \le g'(z) \le 0. \tag{40}$$

Integrating from z_1 to z_2 we find

$$-\frac{1}{12}z_1^{-1} \le -\frac{1}{12}(z_1^{-1} - z_2^{-1}) \le g(z_2) - g(z_1) \le 0. \tag{41}$$

Thus

$$C = \lim_{z \to \infty} g(z) \tag{42}$$

exists by Cauchy's criterion and

$$0 \le g(z) - C \le \frac{1}{12}z^{-1}. \tag{43}$$

Now we evaluate the constants c and C. From equation (39) and the functional equation of the Gamma function, we see that

$$g(z + 1) - g(z) = -(z + \frac{1}{2}) \log \frac{z + 1}{z} + 1 - c. \tag{44}$$

Taylor's theorem applied to the logarithm gives

$$\lim_{z \to \infty} (z + \frac{1}{2}) \log \frac{z+1}{z} = 1 \tag{45}$$

and hence

$$\lim_{z \to \infty} [g(z+1) - g(z)] = c. \tag{46}$$

From (41) we conclude that

$$c = 0. \tag{47}$$

Using equation (39) again we see that

$$g(2z) - g(z) - g(z + \frac{1}{2}) = \log \frac{2^{\frac{1}{2}-2z}\Gamma(2z)}{\Gamma(z)\Gamma(z + \frac{1}{2})} - z \log \frac{z + 1/2}{z} + \frac{1}{2}. \tag{48}$$

Taylor's theorem gives

$$\lim_{z \to \infty} z \log \frac{z + 1/2}{z} = \frac{1}{2} \tag{49}$$

while the Legendre duplication formula gives

$$2^{\frac{1}{2}-2z} \frac{\Gamma(2z)}{\Gamma(z)\Gamma(z + \frac{1}{2})} = (2\pi)^{-1/2}. \tag{50}$$

Thus we see that

$$\lim_{z \to \infty} [g(2z) - g(z) - g(z + \frac{1}{2})] = -\frac{1}{2} \log 2\pi. \tag{51}$$

On the other hand

$$\lim_{z \to \infty} [g(2z) - g(z) - g(z + \frac{1}{2})] = \lim_{z \to \infty} g(2z) - \lim_{z \to \infty} g(z) - \lim_{z \to \infty} g(z + \frac{1}{2}) = -C, \tag{52}$$

so

$$C = \frac{1}{2} \log 2\pi . \tag{53}$$

Equation (43) says that

$$0 \le \log \Gamma(z) - (z - \frac{1}{2}) \log z + z - \frac{1}{2} \log 2\pi \le \frac{1}{12} z^{-1}. \tag{54}$$

Exponentiating we obtain the estimates (3).

Exercises

1. Prove (6) by the following alternate method. Show that the quantity on the right satisfies the functional equation, special value, and log convexity and conclude from the uniqueness theorem that it must be $\Gamma(z)$.

2. Find an alternate method to evaluate the constant C based on the estimates on $\binom{2n}{n}$ derived in Section 1.1.

3. Prove the bounds (4) for $z > 0$. For which z are these bounds sharper than (3)?

4. A "perfect" deal in the game of Bridge is one in which each of the four players is dealt cards of only one suit. With a properly shuffled deck it should occur, on average, once in every

$$\frac{52!}{4!(13!)^4}$$

deals. There are occasional newspaper reports of perfect hands. Find an upper bound on the probability of a perfect hand. For 52! and 13! the estimates (4) are quite adequate. For 4! the obvious formula

$$4! = 1 \cdot 2 \cdot 3 \cdot 4 = 24$$

is simpler and more accurate.

1.4   The Beta Integral

Our next integral, also due to Euler, is

$$I(p, q) = \int_0^\infty t^{p-1}(1 + t)^{-p-q}\, dt, \tag{1}$$

convergent for $p, q > 0$. Our investigation of this integral will follow the same lines as that of the Gamma integral. We begin by integrating by parts to establish the functional equation,

$$pI(p, q) = (p + q)I(p + 1, q). \tag{2}$$

In more detail,

$$
\begin{aligned}
p \int_0^\infty t^{p-1}(1+t)^{-p-q}\, dt &= \int_0^\infty \frac{d}{dt} t^p (1+t)^{-p-q}\, dt \tag{3}\\
&= -\int_0^\infty t^p \frac{d}{dt}(1+t)^{-p-q}\, dt\\
&= (p+q) \int_0^\infty t^p (1+t)^{-p-q-1}\, dt.
\end{aligned}
$$

We can prove another functional equation,

$$I(p, q) = I(p, q + 1) + I(p + 1, q) \tag{4}$$

by a trick which has no counterpart for the Gamma integral. We can expand one factor of $(1 + t)$. In other words, we write

$$(1 + t)^{-p-q} = (1 + t)^{-p-q-1} + t(1 + t)^{-p-q-1}, \tag{5}$$

and integrate both sides against $t^{p-1} dt$,

$$\int_0^\infty t^{p-1}(1 + t)^{-p-q} \, dt = \int_0^\infty t^{p-1}(1 + t)^{-p-q-1} \, dt + \int_0^\infty t^p(1 + t)^{-p-q-1} \, dt.$$

Again there are certain values of the arguments for which the integral is easy to evaluate, for example

$$I(1, q) = \frac{1}{q}. \tag{6}$$

We also have a log convexity inequality, proved by taking

$$f(t) = t^{p-1}(1 + t)^{-p-q}$$
$$g(t) = t^{p'-1}(1 + t)^{-p'-q'}$$

in Hölder's inequality. This gives

$$I(\alpha p + \alpha' p', \alpha q + \alpha' q') \leq I(p, q)^\alpha I(p', q')^{\alpha'} \tag{7}$$

for $\alpha, \alpha', p, p', q, q' > 0$ and $\alpha + \alpha' = 1$. The relations (3), (4), (6), and (7) determine I completely. That is the content of the following theorem.

THEOREM 3 (UNIQUENESS OF BETA)
There is one and only one function $B(p, q)$ with the following properties.

1. *Functional equations,*

$$pB(p, q) = (p + q)B(p + 1, q) \tag{8}$$

and

$$B(p, q) = B(p, q + 1) + B(p + 1, q). \tag{9}$$

2. *Special value,*

$$B(1, q) = \frac{1}{q}. \tag{10}$$

3. Log convexity,

$$B(\alpha p + \alpha' p', \alpha q + \alpha' q') \leq B(p, q)^\alpha B(p', q')^{\alpha'} \tag{11}$$

for all $\alpha, \alpha', p, p', q, q' > 0$ *with* $\alpha + \alpha' = 1$.

Proof. We have already established existence, so we proceed to uniqueness. Let B be any function with these three properties.

First we prove an improved version of the first functional equation,

$$(p)_m B(p, q) = (p + q)_m B(p + m, q) \tag{12}$$

by induction. Taking $p = 1$ and using the special value, we find

$$B(n + 1, q) = \frac{n!}{(q)_{n+1}}. \tag{13}$$

Next we prove the limit formula

$$\lim_{u \to \infty} \frac{B(c + u, q)}{B(b + u, q)} = 1. \tag{14}$$

As with the Gamma function we assume, without loss of generality, that $b < c$ and find a, d such that $a < b < c < d$ and the differences $c - a$ and $d - b$ are both integers. Log convexity and the functional equation give the estimates

$$\left(\frac{(u + a)_{c-a}}{(u + a + q)_{c-a}} \right)^{\frac{b-a}{c-a}} \leq \frac{B(c + u, q)}{B(b + u, q)} \leq \left(\frac{(u + b)_{d-b}}{(u + b + q)_{d-b}} \right)^{\frac{c-b}{d-b}}. \tag{15}$$

The upper and lower bounds both tend to 1 as u tends to infinity so the limit formula (14) is proved.

Taking $b = 1$, $c = p + 1$, and restricting u to integers n, we get

$$\lim_{n \to \infty} \frac{B(p + n + 1, q)}{B(n + 1, q)} = 1 \tag{16}$$

or, using (12) and (13),

$$B(p, q) = \lim_{n \to \infty} \frac{n!(p + q)_{n+1}}{(p)_{n+1}(q)_{n+1}}. \tag{17}$$

This formula determines $B(p, q)$ uniquely, so the theorem is proved.

Several observations are worth making. The limit on the right in (17) is well-defined and finite if neither p nor q is a nonpositive integer. We now take this representation as the definition of the Beta function. The integral representation (1) is then a formula valid only under the restriction $p, q > 0$.

Since we never needed the functional equation (9), the theorem shows that it must follow from (8), (10), and (11), though this is far from obvious. More interesting still, the representation (17) shows that

$$B(p, q) = B(q, p), \tag{18}$$

which is not obvious from the integral representation (1). It can be proved by a change of variable. Setting $s = t^{-1}$ in (1) we get

$$I(p, q) = \int_0^\infty s^{q-1}(1 + s)^{-p-q} \, ds = I(q, p). \tag{19}$$

Most importantly, by comparing (17) with equation (34) of Section 1.1, we see that

$$B(p, q) = \frac{\Gamma(p)\Gamma(q)}{\Gamma(p+q)}. \tag{20}$$

Exercises

1. The Beta integral has several other integral representations, also due to Euler, for example

$$B(p, q) = \int_0^1 s^{p-1}(1 - s)^{q-1}, \, ds \tag{21}$$

and

$$B(p, q) = 2 \int_0^{\pi/2} \sin^{2p-1} \theta \cos^{2q-1} \theta \, d\theta, \tag{22}$$

for $p, q > 0$. Prove these by finding an appropriate change of variables in the integral (1).

2. Strictly speaking, we proved the functional equation (4) using the integral representation, so we can only be sure that it holds when $p, q > 0$. Use the limit representation (17) to show that it holds for all p, q for which $B(p, q)$ is defined.

3. Use Taylor's theorem and the integral representation (21) to show that

$$B(p, q) = \sum_{n=0}^\infty \frac{(1 - q)_n}{n!(p+q+n)}. \tag{23}$$

4. Express π in terms of $B(n + \frac{1}{2}, n + \frac{1}{2})$ and $\binom{2n}{n}$.

5. Log convexity gives the bounds

$$B(n + \frac{1}{2}, n + \frac{1}{2}) \leq B(n, n + 1).$$

Use the formula you obtained in the preceding exercise to obtain a sequence of upper bounds for π. These bounds were first discovered by Wallis in [69].

6. Does the sequence you found in the preceding exercise converge to π? If so, how fast?

1.5   The Whittaker Function

Our next integral, the Laplace transform of $t^{p-1}(1 + t)^{-p-q}$,

$$I(p, q, s) = \int_0^\infty t^{p-1}(1 + t)^{-p-q} e^{-st}\, dt, \tag{1}$$

converges locally uniformly in p, q, and s for $p, s > 0$ or $s = 0$ and $p, q > 0$. It was first introduced by Dirichlet[5] in [16]. The first systematic investigation was Kummer's article [46]. It is a sort of hybrid between the Gamma and Beta integrals, and it can be studied by the same methods. Many interesting functions can be written in terms of the Whittaker function. A partial list will be given at the end of the section.

The integral $I(p, q, s)$ satisfies the functional equations

$$pI(p, q, s) = (p + q)I(p + 1, q, s) + sI(p + 1, q - 1, s), \tag{2}$$

and

$$I(p, q, s) = I(p, q + 1, s) + I(p + 1, q, s), \tag{3}$$

proved, respectively, by integration by parts and by integrating the identity

$$(1 + t)^{-p-q} = (1 + t)^{-p-q-1} + t(1 + t)^{-p-q-1} \tag{4}$$

against $t^{p-1}e^{-st}\, dt$. We can find special values by choosing p, q, s such that (1) becomes either a Beta or Gamma integral,

$$I(p, q, 0) = B(p, q), \tag{5}$$

$$I(p, -p, s) = \Gamma(p)s^{-p}. \tag{6}$$

---

[5]This article appears to have achieved complete oblivion. The representation (1) is universally attributed to Whittaker, whose paper [76] was written 68 years later.

Applying Hölder's inequality with

$$f(t) = t^{p-1}(1+t)^{-p-q}e^{-st},$$
$$g(t) = t^{p'-1}(1+t)^{-p'-q'}e^{-s't},$$

we obtain the log convexity inequality

$$I(\alpha p + \alpha' p', \alpha q + \alpha' q', \alpha s + \alpha' s') \leq I(p, q, s)^\alpha I(p', q', s')^{\alpha'} \tag{7}$$

where $\alpha, \alpha' > 0$ and $\alpha + \alpha' = 1$. This inequality is, however, much less useful than the log convexity inequalities for the Gamma and Beta functions.

If we try to prove a limit formula similar to those for the Gamma and Beta integrals, we find that the functional equations (2) and (3) are not especially helpful.[6] The problem is that these equations relate the value of I at *three* different points, rather than just two. There are countably many similar functional equations, all of the general form

$$c_1 I(p + h_1, q + k_1, s) + c_2 I(p + h_2, q + k_2, s) + c_3 I(p + h_3, q + k_3, s) = 0. \tag{8}$$

Here $h_1, k_1, h_2, k_2, h_3, k_3$ are all integers and c_1, c_2, c_3 are polynomials in the variables p, q, s with integer coefficients. In fact, there is such a relation whenever the three pairs (h_1, k_1), (h_2, k_2), and (h_3, k_3) are distinct. As an example we will prove the three-term recurrence

$$(p + q)I(p + 1, q, s) - (2p + q + s - 1)I(p, q, s)$$
$$+ (p - 1)I(p - 1, q, s) = 0. \tag{9}$$

A useful bookkeeping device is to represent a relation of the form (8) by a triangle with vertices at the points (h_1, k_1), (h_2, k_2), and (h_3, k_3). In the diagram the shaded triangles represent relations which may be obtained from the relation (3) by adding appropriate constants to p and q. For example, the triangle BCE corresponds to the relation

$$I(p - 1, q, s) = I(p - 1, q + 1, s) + I(p, q, s). \tag{10}$$

The unshaded triangles represent relations which may be obtained from the relation (2) by adding appropriate constants to p and q. For example, the triangle BDE corresponds to the relation

$$(p - 1)I(p - 1, q + 1, s) = (p + q)I(p, q + 1, s) + sI(p, q, s). \tag{11}$$

---

[6]There is, in some sense an analogue for the Whittaker integral of the limit representations of the Gamma and Beta integrals. It is not a limit of products but a continued fraction expansion. This can be proved using the three term recurrence derived below, but we will not do so.

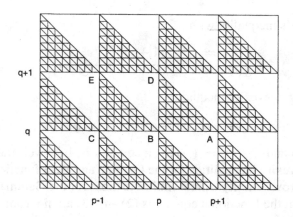

Figure 1.1: Geometric aid for calculation of three term recurrences.

The degenerate triangle ABC represents the relation (9), which we wish to prove.

We proceed as follows. By eliminating the $I(p, q + 1, s)$ term, which corresponds to the vertex D, from the relations (3) and (11), which correspond to the triangles ABD and BDE, we obtain the relation

$$(p - 1)I(p - 1, q + 1, s) = (p + q + s)I(p, q, s) - (p + q)I(p + 1, q, s), \tag{12}$$

which corresponds to the triangle ABE. Eliminating the $I(p - 1, q + 1, s)$ term, which corresponds to the vertex E, from this relation and the relation (10), which corresponds to the triangle BCE, we obtain (9).

The three term recurrence (9) has many uses. In a moment we will use it to derive the differential equation satisfied by $I(p, q, s)$. Now we use it to extend the function beyond the region $p > 0$ where it is initially defined. The general idea is quite simple. Writing (9) in the form

$$I(p, q, s) = -\frac{p + q + 1}{p}I(p + 2, q, s) + \frac{2p + q + s + 1}{p}I(p + 1, q, s), \tag{13}$$

we note that the right hand side makes sense when $-1 < p < 0$. We can therefore use equation (13) to define $I(p, q, s)$ for these values of p. Having done so, we find that the right hand side now makes sense when $-2 < p < -1$. We could continue this procedure indefinitely, gradually extending $I(p, q, s)$ to all values of p except the nonpositive integers. When we consider the Laguerre polynomials, however, we will be interested in precisely the case where p is a nonpositive integer. While we cannot extend $I(p, q, s)$ to these points in a continuous way, we can extend the function

$$J(p, q, s) = \Gamma(p)^{-1}I(p, q, s). \tag{14}$$

We therefore modify the argument sketched above in such a way that it applies to the function $J(p, q, s)$. In place of (13) we now have

$$J(p, q, s) = -(p + q + 1)J(p + 2, q, s) \tag{15}$$
$$+ (p + 1)(2p + q + s + 1)J(p + 1, q, s).$$

Because we have removed the factor of p in the denominator, we now have a right hand side which is well-defined and continuous for $-1 < p$. We therefore use equation (15) to define $J(p, q, s)$ for all these values of p. Having done so, we note that the right hand side is well-defined and continuous for all $-2 < p$. We now use (15) to define $J(p, q, s)$ for these values of p. We continue this process indefinitely, obtaining a function $J(p, q, s)$ which is continuous for all p.

We could have extended the Gamma and Beta integrals by a similar procedure, inductively using their functional equations. The limit representations rendered this argument unnecessary.

Next we consider the behavior in the s variable. The change of variables which we used for the Gamma function gives an integral which is not of the form (1), so this is of no use. Instead we differentiate under the integral sign, which gives the relation

$$\frac{\partial}{\partial s}I(p, q, s) = -I(p + 1, q - 1, s) \tag{16}$$

This is justified by Theorem T of the appendix whenever the integral on the right hand side is convergent locally uniformly in s, and thus whenever $p + 1, s > 0$. Using the relation (2) we can rewrite (16) as

$$(s\frac{\partial}{\partial s} + p)I(p, q, s) = (p + q)I(p + 1, q, s). \tag{17}$$

This allows us to interpret the three term recurrence (9) as a second order linear differential equation

$$\{s\frac{\partial^2}{\partial s^2} - (q + s - 1)\frac{\partial}{\partial s} - p\}I(p, q, s) = 0. \tag{18}$$

$J(p, q, s)$, since it differs only by a factor independent of s, must satisfy the same equation.

If we expand the factor of e^{-st} in the integral (1) in Taylor series and integrate term by term, we find

$$I(p, q, s) \overset{?}{=} \sum_{l=0}^{\infty} \frac{(-s)^l}{l!} \int_0^{\infty} t^{p+l-1}(1+t)^{-p-q}\, dt$$

$$\overset{?}{=} \sum_{l=0}^{\infty} \frac{(-s)^l}{l!} B(p+l, q-l)$$

$$\overset{?}{=} B(p, q) \sum_{l=0}^{\infty} \frac{(p)_l (s)^l}{(1)_l (1-q)_l}$$

where the $\overset{?}{=}$ sign indicates that we have made no attempt to justify the term by term integration of the series or replacing the Beta integral by the Beta function when the former does not converge. Nonetheless the series we have obtained is easily seen to converge for all s. Term by term differentiation is justified by Theorem O of the appendix, and we can check that the series satisfies the same differential equation, (18), as $I(p, q, s)$. It is easy to check that the values and first derivatives at $s = 0$ agree, at least for those values of q for which the derivative exists, so we may appeal to the existence and uniqueness theorem for differential equations to turn the $\overset{?}{=}$ into an $=$.

In view of the preceding remarks, it may come as a surprise that

$$I(p, q, s) \neq B(p, q) \sum_{l=0}^{\infty} \frac{(p)_l (s)^l}{(1)_l (1-q)_l}.$$

This example illustrates the dangers of performing formal calculations without checking that the hypotheses of the relevant theorems are satisfied. The term by term integration was, of course, invalid. More subtly, the appeal to the existence and uniqueness theorem was spurious. $s = 0$ is a singular point of the differential equation, so the value and first derivative do not suffice to determine the solution uniquely. In fact the correct series expansion is given by equation (32) of Section 2.8. While it is possible to prove this by real variable methods, we will wait and give a complex analytic proof in the next chapter.

We now consider the behavior of $I(p, q, s)$ for s large. Suppose that, undaunted by our recent disaster, we expand $(1 + t)^{-p-q}$ in Taylor series and integrate term by term. This produces

$$I(p, q, s) \overset{?}{=} \Gamma(p) \sum_{n=0}^{\infty} (-1)^n \frac{(p+q)_n (p)_n}{(1)_n} s^{-p-n}$$

The expansion for $J(p, q, s)$ is slightly nicer,

$$J(p, q, s) \overset{?}{=} \sum_{n=0}^{\infty} (-1)^n \frac{(p+q)_n (p)_n}{(1)_n} s^{-p-n}$$

The $\overset{?}{=}$ sign indicates that we have made no attempt to justify the term by term integration. The series on the right is a "formal solution" of the differential equation (18) in the sense that, if we differentiate term by term and group like powers of s without any concern for whether these operations are justified, then we do get zero on the right hand side. The series terminates after only finitely many terms if either p or $p + q$ is a nonpositive integer. In these cases it is possible to show that the formula given above is correct. In *all* other cases the series is divergent for all s. At first sight this is an even greater disaster than that of the preceding paragraph. Nevertheless this calculation, unlike the previous one, can be made to yield useful information.

We repeat the argument, this time with proper attention to questions of convergence. Taylor's theorem with remainder yields the identity

$$(1+t)^{-p-q} = \sum_{n=0}^{m} (-1)^n \frac{(p+q)_n}{(1)_n} t^n + R_m(t), \tag{19}$$

where the remainder R_m has the integral representation

$$R_m(t) = (-1)^{m+1} \frac{(p+q)_{m+1}}{(1)_m} \int_0^t (t-x)^m (1+x)^{-p-q-m-1} \, dx. \tag{20}$$

Integrating against $t^{p-1} e^{-st} \, dt$, we find

$$I(p, q, s) = \Gamma(p) \sum_{n=0}^{m} (-1)^n \frac{(p+q)_n (p)_n s^{-p-n}}{(1)_n} \tag{21}$$

$$+ \int_0^{\infty} R_m(t) t^{p-1} e^{-st} \, dx \, dt.$$

We note that the integrand in (20) is always positive. Once $p + q + m > 0$ the factor $\frac{(p+q)_{m+1}}{(1)_m}$ is of constant sign and $(-1)^m$ is of alternating sign. Thus the sign of $R_m(t)$ is independent of t and, once $p + q + m > 0$, alternating with m. The remaining factors in the integral on the right in (21) are all positive so

$$I(p, q, s) - \Gamma(p) \sum_{n=0}^{m} (-1)^n \frac{(p+q)_n (p)_n s^{-p-n}}{(1)_n}$$

is of alternating sign. In other words, the successive partial sums of the divergent series

$$\Gamma(p)\sum_{n=0}^{\infty}(-1)^n\frac{(p+q)_n(p)_n s^{-p-n}}{(1)_n} \tag{22}$$

furnish upper and lower bounds for $I(p,q,s)$. The usual term for such a series representation is *enveloping*. This implies the weaker property

$$I(p,q,s)=\Gamma(p)\sum_{n=0}^{m-1}(-1)^n\frac{(p+q)_n(p)_n s^{-p-n}}{(1)_n}+O(s^{-p-m}) \tag{23}$$

for s large. In general, a series

$$\sum_{n=0}^{\infty}a_n z^{\alpha-n}$$

is said to be *asymptotic* to function f if

$$f(z)=\sum_{n=0}^{m-1}a_n z^{\alpha-n}+O(z^{\alpha-m}). \tag{24}$$

The first general treatment of such series is Poincaré's paper [58].[7]

The importance of the integral $I(p,q,s)$ stems from the multitude of useful functions which may be written in terms of it. We have already met one of these, the logarithmic integral, in Section 1.2.

$$\operatorname{li} x=\int_0^x\frac{du}{\log u}. \tag{25}$$

There we were concerned with the behavior for x large. For $0<x<1$, the change of variables

$$t=\frac{\log u-\log x}{\log x} \tag{26}$$

converts the integral into a Whittaker integral.

$$\operatorname{li} x=xI(1,0,-\log x). \tag{27}$$

---

[7]A peculiar feature of this paper is that Poincaré was motivated by the example of Stirling's series, as he makes clear in the introduction. He does not, however, prove that Stirling's series is an asymptotic series. He does prove a theorem from which it follows that (22) is an asymptotic series.

Another useful function, particularly in probability and statistics, is the incomplete Gamma function

$$\Gamma(x, z) = \int_{x}^{\infty} u^{z-1} e^{-t} \, dt, \tag{28}$$

first investigated by Legendre. It can be rewritten, using the change of variables $u = \frac{t-x}{x}$, in the form

$$\begin{aligned} \Gamma(x, z) &= x^z e^{-x} \int_{0}^{\infty} (1+t)^{z-1} e^{-xt} \, dt \\ &= x^z e^{-x} I(1, -z, x). \end{aligned} \tag{29}$$

Similarly the error function,

$$\begin{aligned} \text{erf}(z) &= \frac{2}{\sqrt{\pi}} \int_{0}^{z} e^{-u^2} \, du \tag{30} \\ &= 1 - \frac{2}{\sqrt{\pi}} \int_{z}^{\infty} e^{-u^2} \, du, \end{aligned}$$

can be rewritten, using the change of variables $\frac{u^2 - z^2}{z^2}$, as

$$\begin{aligned} \text{erf}(z) &= 1 - \frac{z e^{-z^2}}{\sqrt{\pi}} \int_{0}^{\infty} (1+t)^{-1/2} e^{-z^2 t} \, dt \tag{31} \\ &= 1 - \frac{z e^{-z^2}}{\sqrt{\pi}} I(1, -\tfrac{1}{2}, z^2). \end{aligned}$$

All the functions considered above are defined in terms of $I(p, q, s)$ or, equivalently, $J(p, q, s)$ for $p = 1$. Now we consider the generalized Laguerre polynomials $L_n^{(\alpha)}(x)$, which are defined in terms of $J(p, q, s)$ with p a nonpositive integer. These are given by the formula

$$L_n^{(\alpha)}(x) = \frac{(-1)^n}{n!} J(-n, -\alpha, x) \tag{32}$$

where n is a nonnegative integer. In applications usually α is a nonnegative integer as well, but we do not assume this. $L_n^{(\alpha)}$ is a polynomial of degree n. This would be obvious from the expansion (22). Even though the series is convergent in this case there is, unfortunately, no easy proof using the tools we have developed so far that the series represents $J(p, q, s)$. Proving this from equation (15) and the integral representation for $I(p, q, s)$ is possible, but the computation is horrid. We will simply assume this fact for now. In Section 2.9 we will give a proof.

The differential equation

$$\{x\frac{d^2}{dx^2} + (\alpha - x + 1)\frac{d}{dx} - n\}L_n^{(\alpha)}(x) = 0 \tag{33}$$

and three term recurrence

$$(n+1)L_{n+1}^{(\alpha)}(x) + (x - 2n - \alpha - 1)L_n^{(\alpha)}(x) + (n+\alpha)L_{n-1}^{(\alpha)}(x) = 0 \tag{34}$$

follow immediately from those for $J(p, q, s)$. One of the most important properties of these functions is orthogonality. We say that two functions $y_1(x)$ and $y_2(x)$ on an interval (x_1, x_2) are *orthogonal* with respect to the weight $\mu(x)$ if

$$\int_{x_1}^{x_2} y_1(x)y_2(x)\mu(x)\,dx = 0. \tag{35}$$

For $\alpha > -1$ the Laguerre polynomials $L_m^{(\alpha)}$ and $L_n^{(\alpha)}$ are orthogonal with respect to the weight $x^\alpha e^{-x}$ if $m \neq n$. There are several ways to prove this. We choose to derive it from the differential equation (33). In Section 1.6 we will want to prove the orthogonality of the Jacobi polynomials on a different interval and with respect to a different weight. It would be a pity to give essentially the same argument twice, so we make that part of the argument which applies to arbitrary second order linear ordinary differential equations a separate theorem.

THEOREM 4 (STURM-LIOUVILLE ORTHOGONALITY)
Suppose that y_1, y_2, and u satisfy the differential equations

$$a(x)y_1''(x) + b(x)y_1'(x) + c_1(x)y_1(x) = 0 \tag{36}$$

$$a(x)y_2''(x) + b(x)y_2'(x) + c_2(x)y_2(x) = 0 \tag{37}$$

$$a(x)v'(x) - b(x)v(x) = 0 \tag{38}$$

on the interval (x_1, x_2). Let

$$u(x) = v(x)w(x) \tag{39}$$

where

$$w(x) = y_1(x)y_2'(x) - y_1'(x)y_2(x). \tag{40}$$

If

$$\lim_{x \to x_1^+} u(x) = 0 \tag{41}$$

and

$$\lim_{x \to x_2^-} u(x) = 0 \qquad (42)$$

then y_1 and y_2 are orthogonal on the interval (x_1, x_2) with respect to the weight

$$\mu(x) = \frac{u(x)}{a(x)}(c_2(x) - c_1(x)). \qquad (43)$$

Theorem 4 allows us to prove the orthogonality of two functions by evaluating two limits and solving the differential equation (38). This equation is of first order, so it is usually quite easy to solve explicitly. In most applications the limits are easy to evaluate as well. In our case we have

$$
\begin{aligned}
x_1 &= 0, \qquad\qquad (44)\\
x_2 &= \infty,\\
a(x) &= x,\\
b(x) &= \alpha - x + 1,\\
c_1(x) &= m,\\
c_2(x) &= n,\\
y_1(x) &= L_m^{(\alpha)}(x),\\
y_2(x) &= L_n^{(\alpha)}(x).
\end{aligned}
$$

It is easy to check that

$$v(x) = x^{\alpha+1}e^{-x} \qquad (45)$$

satisfies (38). Both y_1 and y_2 are polynomials, so w is a polynomial. It follows that the limits (42) and (43) are zero. Theorem 4 says that

$$(n - m) \int_0^\infty L_m^{(\alpha)}(x) L_n^{(\alpha)}(x) x^\alpha e^{-x} \, dx = 0. \qquad (46)$$

This is exactly the orthogonality relation we seek. All that remains is to prove Theorem 4.

Proof. From the differential equations (36) and (37) we compute

$$a(x)w'(x) + b(x)w(x) + (c_2(x) - c_1(x))y_1(x)y_2(x) = 0. \qquad (47)$$

Using (38) we see that

$$u'(x) + \mu(x)y_1(x)y_2(x) = 0. \qquad (48)$$

We integrate this equation from ξ_1 to ξ_2,

$$u(\xi_2) - u(\xi_1) + \int_{\xi_1}^{\xi_2} \mu(x)y_1(x)y_2(x) = 0. \tag{49}$$

Now we let ξ_1 tend to x_1 from the right and ξ_2 tend to x_2 from the left. The limits of the first two terms are just the two limits (39) and (40), which we assumed to be zero. The limit of the integral is simply the definition of the (possibly) improper integral (35). The proof of Theorem 4 is thus complete.

It is traditional to use, in place of the functions $I(p, q, s)$ and $J(p, q, s)$ above, the closely related function,

$$W_{\kappa,\mu}(s) = \frac{s^{\mu+1/2}e^{-s/2}}{\Gamma(\mu - \kappa + \frac{1}{2})} \int_0^\infty t^{\mu-\kappa-1/2}(1+t)^{\mu+\kappa-1/2}e^{-st}\,dt. \tag{50}$$

In terms of $J(p, q, s)$ this can be written

$$W_{\kappa,\mu}(s) = s^{\mu+1/2}e^{-s/2}I(\mu - \kappa + \frac{1}{2}, -2\mu, s) \tag{51}$$

or

$$J(p, q, s) = s^{(q-1)/2}e^{s/2}W_{(1-2p-q)/2,-q/2}(s). \tag{52}$$

It is a straightforward, if rather tedious, task to translate properties of $I(p, q, s)$ into properties of $W_{\kappa,\mu}(s)$. For example, the differential equation (18) for $I(p, q, s)$ implies the differential equation

$$\left(s^2\frac{d^2}{ds^2} - \frac{1}{4}s^2 + \kappa s - \mu^2 + \frac{1}{4}\right)W_{\kappa,\mu}(s) = 0 \tag{53}$$

for the Whittaker functions, while the three term recurrence (9) implies the three term recurrence

$$W_{\kappa+1,\mu}(s) + (2\kappa - s)W_{\kappa,\mu}(s) - (\mu - \kappa + \frac{1}{2})(\mu + \kappa - \frac{1}{2})W_{\kappa-1,\mu}(s) = 0 \tag{54}$$

for the Whittaker functions.

The best known and oldest examples of Whittaker functions are the Bessel functions, first introduced by Bessel [3] for a problem in astronomy. The familiar J and Y Bessel functions can be written in terms of values of W on the *imaginary* axis, which we have been systematically ignoring. The K Bessel function, known as the modified Bessel function of the second kind, has the very simple expression

$$K_\nu(x) = \sqrt{\frac{\pi}{2x}}W_{0,\nu}(2x). \tag{55}$$

Exercises

1. Find a three term recurrence relating $I(p-1, q+1, s)$, $I(p, q, s)$, and $I(p+1, q-1, s)$. What is the corresponding recurrence for the Whittaker functions?

2. Prove that the series

$$\sum_{l=0}^{\infty} \frac{(p)_l s^l}{(1)_l (1-q)_l}$$

converges for all s, as claimed.

3. Prove that the series

$$\sum_{l=0}^{\infty} (-1)^n \frac{(p+q)_l (p)_l s^{-p-n}}{(1)_l}$$

diverges for all s, as claimed.

4. Find the enveloping series for the error function. For given s, which pair of successive partial sums gives the best estimates, i.e. those for which the difference between the upper and lower bounds is smallest?

5. The argument which showed that the series (22) is an enveloping series for the integral $I(p, q, s)$ is not specific to the Whittaker integral. Show, more generally, that if

$$\sum_{k=0}^{\infty} a_k t^{k+\alpha}$$

is an enveloping series for the function $f(t)$, then

$$\Gamma(\alpha) \sum_{k=0}^{\infty} (\alpha)_{k+1} a_k s^{-k-\alpha-1}$$

is an enveloping series for its Laplace transform $F(s)$.

6. Hardy's theorem says that if

$$\sum_{k=0}^{\infty} a_k t^{k+\alpha}$$

is an asymptotic series for the function $f(t)$, that is, if

$$f(t) - \sum_{k=0}^{m-1} a_k t^{k+\alpha} = O(t^{m+\alpha})$$

for t sufficiently small, then

$$\Gamma(\alpha) \sum_{k=0}^{\infty} (\alpha)_{k+1} a_k s^{-k-\alpha-1}$$

is an enveloping series for its Laplace transform $F(s)$, that is

$$F(s) - \Gamma(\alpha) \sum_{k=0}^{m-1} (\alpha)_{k+1} a_k s^{-k-\alpha-1} = O(s^{-m-\alpha-1})$$

for s sufficiently large. Prove this theorem.

7. Since we proved the differential equation (53) using the integral representation, it is only known to be valid when the integral converges, i.e. when $\kappa < \mu + \frac{1}{2}$. Use the three term recurrence (54) and induction to show that (53) holds for all κ, μ. This is a very annoying calculation. Later, when we meet the rigidity principle, we will see how to obtain the same result almost effortlessly.

8. What differential equation does the Bessel function K_ν satisfy?

9. What is the behavior at infinity of the Bessel function K_ν?

10. What is $K_{-1/2}(x)$?

11. Suppose $\alpha > -1$ and m and N are nonnegative integers. Prove that

$$\int_0^\infty x^{\alpha+1} e^{-x} L_m^{(\alpha)}(x) L_n^{(\alpha)}(x) \, dx = 0$$

unless $|m - n| \le 1$. *Hint:* Use the three term recurrence and the orthogonality relation.

1.6   The Hypergeometric Function

Our next integral[8]

$$I(a, b, c, z) = \int_0^\infty t^{b-1} (1+t)^{a-c} (1+t-zt)^{-a} \, dt, \tag{1}$$

is locally uniformly convergent for $c > b > 0$ and $z < 1$. It converges also for $c - a > b > 0$ and $z = 1$. Like most interesting functions, it was first studied by Euler, in [27]. Its theory was developed by Gauss [31], Kummer [45], and Thomé [67], among others. At the end of the section we will give a partial list of the functions which may be expressed in terms of the this integral.

---

[8]This is often called Riemann's integral representation. It is certainly older, having appeared in Jacobi's paper [42]. Jacobi attributes the integral to Euler.

The function $I(a, b, c, z)$ satisfies the following three functional equations,

$$bI(a, b, c, z) = (c - a)I(a, b + 1, c + 1, z) + a(1 - z)I(a + 1, b + 1, c + 1, z), \tag{2}$$

$$I(a, b, c, z) = I(a, b, c + 1, z) + I(a, b + 1, c + 1, z), \tag{3}$$

$$I(a, b, c, z) = I(a + 1, b, c, z) - zI(a + 1, b + 1, c + 1, z). \tag{4}$$

Equation (2) is proved by integration by parts, (3) is proved using the familiar identity

$$(1 + t)^{a-c} = (1 + t)^{a-c-1} + t(1 + t)^{a-c-1}, \tag{5}$$

and (4) by the similar identity

$$(1 + t - zt)^{-a} = (1 + t)(1 + t - zt)^{-a-1} - zt(1 + t - zt)^{-a-1}. \tag{6}$$

Many special values are known. When $z = 0$ or $z = 1$ we have a Beta integral,

$$I(a, b, c, 0) = B(b, c - b) = \frac{\Gamma(b)\Gamma(c - b)}{\Gamma(c)}, \tag{7}$$

$$I(a, b, c, 1) = B(b, c - a - b) = \frac{\Gamma(b)\Gamma(c - a - b)}{\Gamma(c - a)}. \tag{8}$$

When $a = 0$ we have another Beta integral,

$$I(0, b, c, z) = B(b, c - b) = \frac{\Gamma(b)\Gamma(c - b)}{\Gamma(c)}. \tag{9}$$

When $a = c$ the changes of variables $u = t - zt$ converts the integral to a Beta integral

$$I(a, b, a, z) = B(b, a - b)(1 - z)^{-b} = \frac{\Gamma(b)\Gamma(a - b)}{\Gamma(a)}(1 - z)^{-b}. \tag{10}$$

More subtly, when $c = 1 + a - b$ and $z = -1$, the quadratic change of variables $u = 4t + t^2$ converts the integral to a Beta integral,

$$I(a, b, 1 + a - b, -1) = 2^{-2b} B(b, \frac{a - 2b + 1}{2}) = 2^{-2b} \frac{\Gamma(b)\Gamma(\frac{a-2b+1}{2})}{\Gamma(\frac{a+1}{2})}, \tag{11}$$

a result of Kummer. There are even cubic changes of variable due to Goursat, but the list above is long enough already.

We have, as usual, a log convexity estimate,

$$I(\alpha a + \alpha' a, \alpha b + \alpha' b', \alpha c + \alpha' c', z) \le I(a, b, c, z)^\alpha I(a', b', c', z')^{\alpha'}, \quad (12)$$

proved using Hölder's inequality. Note that there is no convexity in the z variable.

As with the Whittaker function, the functional equations at our disposal are of little help in proving a limit formula.[9] We can, as in the case of the Whittaker function, prove countably many functional equations of the same general type as (2), (3), and (4). For any triple (h_1, k_1, l_1), (h_2, k_2, l_2), (h_3, k_3, l_3) of triples of integers, there is a relation of the form

$$\sum_{j=1}^{3} c_j I(a + h_j, b + k_j, c + l_j, z) = 0 \quad (13)$$

with polynomials c_1, c_2, c_3 in the variables a, b, c, z with integer coefficients. All of these can be derived from (2), (3), and (4) by the same procedure we used for the Whittaker functions, though the graphical representation is now three-dimensional, and thus difficult to draw.

As an example we prove the three term recurrence

$$C_1 I(a + 1, b + 1, c + 1, z) + C_2 I(a, b, c, z) + C_3 I(a - 1, b - 1, c - 1, z) = 0, \quad (14)$$

where

$$\begin{aligned} C_1 &= az(z - 1) \\ C_2 &= az + bz - c + 1 \\ C_3 &= b - 1. \end{aligned} \quad (15)$$

First we replace a, b, c in (2) by $a - 1, b - 1, c - 1$, obtaining

$$I(a - 1, b - 1, c - 1, z) = I(a, b - 1, c - 1, z) - zI(a, b, c, z). \quad (16)$$

Then we replace a, b, c in (4) by $a, b - 1, c - 1$, which gives

$$\begin{aligned} (b - 1)I(a, b - 1, c - 1, z) &= (c - a - 1)I(a, b, c, z) \\ &+ a(1 - z)I(a + 1, b, c, z). \end{aligned} \quad (17)$$

---

[9]As in the case of the Whittaker function, the best analogue of the limit representations of the Gamma and Beta integrals is a continued fraction expansion, proved using the three term recurrences developed below. For more information consult [42].

Finally we eliminate the terms in $I(a, b-1, c-1, z)$ and $I(a+1, b, c, z)$ from equations (4), (16), and (17) to obtain (14).

The special importance of this recurrence stems from the formula

$$\frac{\partial}{\partial z}I(a, b, c, z) = aI(a+1, b+1, c+1, z), \tag{18}$$

which is easily proved by differentiating under the integral sign in (1). This allows us to rewrite (14) as the differential equation

$$\{z(z-1)\frac{\partial^2}{\partial z^2} + (az + bz + z - c)\frac{\partial}{\partial z} + ab\}I(a, b, c, z) = 0. \tag{19}$$

Another important recurrence,

$$C_1 I(a, b, c+1, z) + C_2 I(a, b, c, z) + C_3 I(a, b, c-1, z) = 0, \tag{20}$$

with

$$\begin{aligned} C_1 &= (c-a)z \\ C_2 &= az + bz - 2cz + z + c - 1 \\ C_3 &= (c-b-1)(z-1), \end{aligned} \tag{21}$$

is left as an exercise.

We can use these three term recurrences to extend

$$\frac{I(a, b, c, z)}{\Gamma(b)\Gamma(c-b)}$$

to a function continuous for all a, b, c, z with $z < 1$. The argument is much the same as that for the Whittaker function. The only complication is that we first remove the restriction $c > b$, by repeated use of (20), and then the restriction $b > 0$, by means of (14).

As with the Whittaker integral, it is traditional to work with a multiple of the integral (1). The hypergeometric function, $F(a, b; c; z)$ is defined by

$$F(a, b; c; z) = \frac{\Gamma(c)}{\Gamma(b)\Gamma(c-b)}I(a, b, c, z), \tag{22}$$

a normalization which has the convenient feature that

$$F(a, b; c; 0) = 1, \tag{23}$$

but the disadvantage that $F(a, b; c; z)$ is undefined when c is a nonpositive integer.

Since the Gamma factors are independent of z, the differential equation is unchanged while the three term recurrences become

$$C_1 F(a+1, b+1; c+1; z) + C_2 F(a-1, b-1; c-1; z)$$
$$+ C_3 F(a, b; c; z) = 0 \quad (24)$$

and

$$C_1' F(a, b; c+1; z) + C_2' F(a, b; c-1; z) + C_3' F(a, b; c; z) = 0. \quad (25)$$

With coefficients

$$\begin{array}{rcl}
C_1 & = & abz(z-1) \\
C_2 & = & c(az+bz-z-c+1) \\
C_3 & = & c(c-1)
\end{array} \quad (26)$$

and

$$\begin{array}{rcl}
C_1' & = & (c-a)(c-b)z \\
C_2' & = & c(az+bz-2cz+z+c-1) \\
C_3' & = & c(c-1)(z-1)
\end{array} \quad (27)$$

respectively.

In order to find a series representation for $I(a, b, c, z)$, we first expand the factor $(1 + t - zt)^{-a}$ in Taylor series in z,

$$(1 + t - zt)^{-a} = \sum_{n=0}^{m} \frac{(a)_n}{(1)_n} t^n (1+t)^{-a-n} z^n + r_m(a, t, z) \quad (28)$$

with remainder

$$r_m(a, t, z) = \frac{(a)_{m+1}}{(1)_m} t^{m+1} \int_0^z (z-s)^m (1+t-zt)^{-a-m-1} \, ds \quad (29)$$

and integrate against $t^{b-1}(1+t)^{a-c} \, dt$ to obtain

$$I(a, b, c, z) = B(b, c-b) \sum_{n=0}^{m} \frac{(a)_n (b)_n}{(1)_n (c)_n} z^n + R_m(a, b, c, z) \quad (30)$$

with remainder

$$R_m(a, b, c, z)$$
$$= \frac{(a)_{m+1}}{(1)_m} \int_0^\infty \int_0^z t^{b+m} (1+t)^{a-c} (z-s)^m (1+t-zt)^{-a-m-1} \, ds \, dt. \quad (31)$$

If $z < 1$ then $(z - s)^m$ is always positive for m even and always negative for m odd. $(a)_{m+1}$ is of constant sign once $m > -a$ and all other factors are positive. It follows that the sign of $R_m(a, b, c, z)$ is alternately positive and negative and thus that the partial sums

$$B(b, c - b) \sum_{n=0}^{m} \frac{(a)_n (b)_n}{(1)_n (c)_n} z^n$$

for successive values of m furnish upper and lower bounds for $I(a, b, c, z)$.

If $|z| < 1$ we can say more. There are two cases to distinguish. Here we treat $a < -1$ and leave $a \geq -1$ as an exercise. Using the two easy estimates

$$\frac{z - s}{1 + t - zt} \leq \frac{z}{1 + t} \tag{32}$$

and

$$1 + t - st \leq 1 + t \tag{33}$$

in (31) we obtain the estimate

$$|R_m(a, b, c, z)| \leq |\frac{(a)_{m+1}}{(1)_m} \int_0^\infty \int_0^z t^{b+m} (1 + t)^{-c-m-1} z^m \, ds \, dt|. \tag{34}$$

The integral on the right is just a Beta integral, so we find

$$|R_m(a, b, c, z)| \leq (1 + \frac{1}{m}) B(b, c - b) \frac{(a)_{m+1} (b)_{m+1}}{(1)_{m+1} (c)_{m+1}} z^m|. \tag{35}$$

By the ratio test, this tends to zero as m tends to infinity, so

$$I(a, b, c, z) = B(b, c - b) \sum_{n=0}^{\infty} \frac{(a)_n (b)_n}{(1)_n (c)_n} z^n \tag{36}$$

and hence

$$F(a, b; c; z) = \sum_{n=0}^{\infty} \frac{(a)_n (b)_n}{(1)_n (c)_n} z^n \tag{37}$$

for $|z| < 1$.

Now we consider two classes of identities quite unlike those we derived for the Gamma, Beta, and Whittaker functions. The first of these, due to Euler, can be proved by changing variables in the integral (1),

$$\begin{aligned}
I(a, b, c, z) &= (1 - z)^{c-a-b} I(c - a, c - b, c, z) \tag{38} \\
&= (1 - z)^{-b} I(c - a, b, c, \frac{z}{z - 1}) \\
&= (1 - z)^{-a} I(a, c - b, c, \frac{z}{z - 1}).
\end{aligned}$$

integrals. These are due to Thomé.[10] An example is

$$F(a, b; c; x) = \frac{\Gamma(c)\Gamma(b-a)}{\Gamma(b)\Gamma(c-a)}(-x)^{-a}F(a, 1+a-c; 1+a-b; x^{-1}) \quad (39)$$

$$+ \frac{\Gamma(c)\Gamma(a-b)}{\Gamma(a)\Gamma(c-b)}(-x)^{-b}F(b, 1+b-c; 1+b-a; x^{-1}),$$

which we will prove in Section 2.9. We could prove this now using the existence and uniqueness theorem for differential equations. This is not an easy matter since the only points where we know the values of $I(a, b, c, z)$ are $z = 0$ and $z = 1$, where the hypotheses of the existence and uniqueness theorem are violated. A similar problem led to disaster in the preceding section. To avoid this, we would therefore need to approximate $I(a, b, c, z)$ at $z = \epsilon$ and $z = 1 - \epsilon$ with a small positive ϵ, apply the theorem, and then let ϵ tend to 0. Since there are fifteen further identities of the same general type as (39), we should be very unhappy to have to carry this argument through. Fortunately there is a better strategy, though we will need to wait for the techniques of the next chapter.

There are many interesting functions which can be expressed in terms of the integrals $I(a, b, c, z)$. Among the most important are the complete elliptic integrals,

$$K(\kappa) = \int_0^1 \frac{du}{\sqrt{(1-u^2)(1-\kappa^2 u^2)}} \quad (40)$$

$$E(\kappa) = \int_0^1 \sqrt{\frac{(1-\kappa^2 u^2)}{(1-u^2)}}\, du.$$

These play an important role in the theory of elliptic functions, which we will discuss in chapter 3. These integrals can be written in the form (1) by means of the change of variables $t = \frac{u^2}{1-u^2}$,

$$K(\kappa) = \frac{1}{2}I\left(\frac{1}{2}, \frac{1}{2}, 1, \kappa^2\right) \quad (41)$$

$$E(\kappa) = \frac{1}{2}I\left(-\frac{1}{2}, \frac{1}{2}, 1, \kappa^2\right)$$

---

[10]Few mathematicians show much regard for historical accuracy, but later writers have been particularly negligent concerning Thomé. His identities are variously attributed to Kummer, Riemann, and Goursat. Only Barnes seems to have taken the trouble trace them back to their source. Unfortunately the same volume of Crelle which contains Thomé's paper [67] has a paper by Thomae, also on the hypergeometric series. Barnes confuses the two and attributes Thomé's work to Thomae. To our knowledge there is *no* correct reference to Thomé anywhere in the literature.

We may also consider the Whittaker functions as limiting cases of the hypergeometric function. Using the familiar identity

$$e^{-st} = \lim_{n \to \infty} (1 + \frac{st}{n})^{-n} \tag{42}$$

in the Whittaker integral

$$\int_0^\infty t^{p-1}(1+t)^{-p-q}e^{-st}\,dt, \tag{43}$$

we find

$$\int_0^\infty t^{p-1}(1+t)^{-p-q}e^{-st}\,dt = \lim_{n \to \infty} I(n, p, p+q+n, 1 - \frac{s}{n}). \tag{44}$$

For this reason the Whittaker functions are often called confluent hypergeometric functions.

If a is a negative integer, then the power series (36) terminates after finitely many terms and $F(a, b; c; z)$ is a polynomial. The polynomials which arise in this way are called Jacobi polynomials. They were first studied in [40], although special cases were known earlier. The usual notation for the Jacobi polynomials is

$$P_n^{(\alpha,\beta)}(x) = \frac{(\alpha + 1)_n}{(1)_n} F(a, b; c; z), \tag{45}$$

where

$$
\begin{aligned}
a &= -n \\
b &= \alpha + \beta + n + 1 \\
c &= \alpha + 1 \\
z &= \frac{1}{2}(1 - x)
\end{aligned}
\tag{46}
$$

or

$$
\begin{aligned}
\alpha &= c - 1 \\
\beta &= a + b - c \\
n &= -a \\
x &= 1 - 2z.
\end{aligned}
\tag{47}
$$

These include, as special cases, the Legendre, Chebyshev, and ultraspherical polynomials. It is a simple matter to derive the three term recurrence and differential equation for the Jacobi polynomials from those for the hypergeometric function.

The most important property of the Chebyshev polynomials is their orthogonality on the interval $(-1, 1)$ with respect to the weight $(1 - x)^\alpha (1 + x)^\beta$,

$$\int_{-1}^{1} (1 - x)^\alpha (1 + x)^\beta P_m^{(\alpha,\beta)}(x) P_n^{(\alpha,\beta)}(x)\, dx = 0, \qquad (48)$$

if $\alpha, \beta > -1$ and $m \neq n$. We prove this using Theorem 4. Here

$$
\begin{aligned}
x_1 &= -1, && (49)\\
x_2 &= 1,\\
a(x) &= x^2 - 1\\
b(x) &= \alpha x + \beta x + 2x + \alpha - \beta,\\
c_1(x) &= m(m + \alpha + \beta + 1),\\
c_2(x) &= n(n + \alpha + \beta + 1),\\
y_1(x) &= P_m^{(\alpha,\beta)}(x),\\
y_2(x) &= P_n^{(\alpha,\beta)}(x).
\end{aligned}
$$

The function

$$v(x) = (1 - x)^{\alpha+1}(1 + x)^{\beta+1} \qquad (50)$$

satisfies the required differential equation. Theorem 4 therefore implies

$$(m - n)(m + n + \alpha + \beta + 1) \int_{-1}^{1} (1 - x)^\alpha (1 + x)^\beta P_m^{(\alpha,\beta)}(x) P_n^{(\alpha,\beta)}(x)\, dx = 0, \qquad (51)$$

Since, by assumption, $\alpha, \beta > -1$, we have

$$m + n + \alpha + \beta + 1 > m + n - 1. \qquad (52)$$

Since m and n are nonnegative integers

$$m + n + \alpha + \beta + 1 \neq 0, \qquad (53)$$

except possibly when $m = n = 0$. Equation (51) therefore implies (48).

Exercises

1. Prove the three term recurrence (20).

2. Write the incomplete Beta integral,

$$B(p, q, x) = \int_0^x t^{p-1}(1-t)^{q-1}\, dt$$

in terms of the hypergeometric function.

3. The oldest and most interesting of the Jacobi polynomials are the Legendre polynomials, first investigated by Legendre in [49]. They arise in contexts as diverse as numerical integration and the arithmetic of elliptic curves. They can be obtained by taking $\alpha = \beta = 0$ in (46). There are several other ways to define the Legendre polynomials. One is from the generating function,

$$f(x, t) = (1 - 2xt + t^2)^{-1/2}.$$

In terms of f,

$$P_n(x) = \frac{1}{n!}\frac{\partial^n f}{\partial t^n}.$$

Use this representation to compute P_0 through P_5.

4. The function f satisfies the partial differential equation

$$(1 - 2xt + t^2)\frac{\partial f}{\partial t} = (x - t)f.$$

What does this imply about the Legendre polynomials?

5. Prove the elementary integral

$$\int (ax^2 + bx + c)^{-1/2}\, dx = a^{-1/2}\log(2\sqrt{a(ax^2 + bx + c)} + 2ax + b)$$
$$- \frac{a^{-1/2}}{2}\log(4ac - b^2).$$

Apply this to

$$\int_{-1}^1 (1 - 2xs + s^2)^{-1/2}(1 - 2xt + t^2)^{-1/2}\, dx.$$

What does this imply about the Legendre polynomials?

6. Show that equation (36) holds also for $a \geq -1$ and $|z| < 1$.

7. Jacobi proves the differential equation for the hypergeometric function as follows. Under the change of variable $\tau = \frac{t}{t+1}$ equation (1) becomes

$$I(a, b, c, z) = \int_0^1 v(\tau, z)\, d\tau,$$

where

$$v(\tau, z) = \tau^{b-1}(1 - \tau)^{c-b-1}(1 - \tau z)^{-a}.$$

An easy calculation shows that

$$\{z(z - 1)\frac{\partial^2}{\partial z^2} + (az + bz + z - c)\frac{\partial}{\partial z} + ab\}v(\tau, z) = a\frac{\partial}{\partial \tau}\{\frac{\tau(1 - \tau)}{1 - \tau z}v(\tau, z)\}.$$

Applying this to the integral representation above we find

$$\{z(z - 1)\frac{\partial^2}{\partial z^2} + (az + bz + z - c)\frac{\partial}{\partial z} + ab\}I(a, b, c, z) = a[\frac{\tau(1 - \tau)}{1 - \tau z}v(\tau, z)]_{\tau=0}^{\tau=1}.$$

The right hand side is zero for $c < b < 0$. Jacobi then shows that the same argument applies to integrals of the form

$$\int_x^y v(\tau, z)\, d\tau,$$

where $x = -\infty$, $y = 0$ or $x = 1$, $y = +\infty$. Under what conditions do these integrals converge and under what conditions are they solutions of the hypergeometric differential equation? Find appropriate changes of variable to write these integrals in terms of I.

8. Suppose $\alpha, \beta > -1$ and m and N are nonnegative integers. Prove that

$$\int_{-1}^{1} (1 - x)^{\alpha+1}(1 + x)^{\beta+1}x P_m^{(\alpha,\beta)}(x) P_n^{(\alpha,\beta)}(x)$$

unless $|m - n| \leq 1$. Hint: Use the three term recurrence and the orthogonality relation.

9. The hydrogen orbitals are the functions

$$\psi_{nlm}(\rho, \theta, \phi) = -Cr^l e^{-r/2} L_{n+l}^{(2l+1)}(r) P_l^{(|m|,0)}(\cos\theta)e^{im\phi}$$

where

$$r = \frac{8\pi^2 m_0 \epsilon^2}{nh^2}\rho$$

and

$$C = \sqrt{(\frac{8\pi^2 m_0 \epsilon^2}{hh^2})^3 \frac{(n - l - 1)!(l - |m|)!}{(n + l)!^3(l + |m|)!} \frac{2l + 1}{4n}}.$$

Here h, m_0 and ϵ are physical constants whose values depend on the choice of units. l, m and n are integers satisfying

$$0 < n$$
$$0 < l < n$$
$$-l < m < l$$

Compute the hydrogen orbitals for small values of n.

1.7 Euler-MacLaurin Summation

The Euler-MacLaurin formula, in its original form, reduces the computation of a finite sum to that of a definite integral and an infinite sum. This may not sound like a reduction. We will soon see, however, that the formula is more useful than this description would suggest. Its main uses are to replace slowly convergent series by rapidly convergent series and to replace divergent series by convergent series.

There are several nearly equivalent forms of the formula. Euler and MacLaurin did not state, but would certainly have recognized, the formula which now bears their names in the following form:

$$\sum_{j=1}^{n} f(j) = \int_{0}^{n} f(x)\, dx + \sum_{k=1}^{\infty} (-1)^k \frac{B_k}{k!} [f^{(k-1)}(n) - f^{(k-1)}(0)]. \qquad (1)$$

For Euler's exact formulation, see [25]. For MacLaurin's, see [52].

The status of (1) is similar to that of Taylor's form of Taylor's theorem, equation (16) of the appendix. It is not true without some restrictions on f. Much more useful for practical purposes are the corresponding formulae with the sum over k truncated at some value m, and the resulting error written as an integral. In the case of the Euler-MacLaurin formula, this was first accomplished by Jacobi in his paper [39]. There is one crucial difference between (1) and Taylor's form of Taylor's theorem. While we will show in Section 2.4 that the latter is true, at least locally, for arbitrary analytic functions, the former is true only under much stronger hypotheses. For this reason the formula of Euler and MacLaurin is almost never used in the form (1).

In order to state Jacobi's theorem, we need to introduce the *Bernoulli polynomials* $B_k(x)$. We define

$$B_k(x) = \frac{\partial^k u}{\partial t^k}\Big|_{t=0} \qquad (2)$$

where

$$u(x, t) = \begin{cases} 1 & \text{if } t = 0, \\ \frac{t e^{tx}}{e^t - 1} & \text{if } t \neq 0. \end{cases} \qquad (3)$$

If u has an expansion in powers of t then, by Taylor's theorem, that expansion can only be

$$u(x, t) = \sum_{k=0}^{\infty} \frac{B_k(x)}{k!} t^k. \qquad (4)$$

In Section 2.7 we will show that equation (4) holds for $|t| < 2\pi$, but we have no need of this fact in this section. The *Bernoulli numbers* B_k are defined by $B_k = B_k(0)$.

It is a simple matter to translate properties of the function u into properties of the Bernoulli polynomials. Differentiating the identity

$$u(1, t) - u(0, t) = t \tag{5}$$

k times with respect to t, we find that

$$B_k(1) - B_k(0) = \begin{cases} 1 & \text{if } k = 1, \\ 0 & \text{if } k \neq 1. \end{cases} \tag{6}$$

Differentiating

$$u(1 - x, t) = u(x, -t) \tag{7}$$

produces

$$B_k(1 - x) = (-1)^k B_k(x). \tag{8}$$

Using

$$\frac{\partial u}{\partial x} = tu, \tag{9}$$

we prove by induction that

$$\frac{\partial^{k+1} u}{\partial x \partial t^k} = k \frac{\partial^{k-1} u}{\partial t^{k-1}} + t \frac{\partial^k u}{\partial t^k}, \tag{10}$$

which shows that

$$B_k'(x) = k B_{k-1}(x) \tag{11}$$

for $k > 0$. In particular, since

$$B_0(x) = 1, \tag{12}$$

it follows that $B_k(x)$ is a polynomial of degree k.

We can now state Jacobi's theorem. The actual statement in [39] is slightly different, but the content is the same.

THEOREM 5 (EULER-MACLAURIN-JACOBI)
If f is m times continuously differentiable on the interval $[0, n]$, *then*

$$\sum_{j=1}^{n} f(j) = \int_{0}^{n} f(x)\,dx + \sum_{k=1}^{m} (-1)^k \frac{B_k}{k!} [f^{(k-1)}(n) - f^{(k-1)}(0)] \quad (13)$$

$$+ (-1)^{m+1} \sum_{j=1}^{n} \int_{j-1}^{j} \frac{B_m(x-j+1)}{m!} f^{(m)}(x)\,dx.$$

This theorem would not be of much use if one had to evaluate the error term

$$E = \sum_{j=1}^{n} \int_{j-1}^{j} \frac{B_m(x-j+1)}{m!} f^{(m)}(x)\,dx.$$

In practice, one normally estimates the integrals, for example by

$$|E| \le \max_{0 \le x \le 1} \frac{|B_m(x)|}{m!} \int_{0}^{n} |f^{(m)}(x)|\,dx, \quad (14)$$

and then chooses m, n, and any other available parameters, so as to make the right hand side small. We will see in Section 2.7 that

$$\max_{0 \le x \le 1} |B_m(x)| = B_m \quad (15)$$

for m even.

Proof. From (12) we see that

$$\int_{j-1}^{j} f(x)\,dx = \int_{j-1}^{j} \frac{B_0(x-j+1)}{0!} f(x)\,dx. \quad (16)$$

The identity (11) makes integration by parts quite easy,

$$\int_{j-1}^{j} \frac{B_{k-1}(x-j+1)}{(k-1)!} f^{(k-1)}(x)\,dx = \int_{j-1}^{j} \frac{d}{dx} \left(\frac{B_k(x-j+1)}{k!} \right) f^{(k-1)}(x)\,dx \quad (17)$$

$$= \frac{B_k(1)}{k!} f^{(k-1)}(j)$$

$$- \frac{B_k(0)}{k!} f^{(k-1)}(j-1)$$

$$- \int_{j-1}^{j} \frac{B_k(x-j+1)}{k!} f^{(k)}(x)\,dx.$$

We apply this identity repeatedly to (16) and obtain

$$\int_{j-1}^{j} f(x)\,dx = \sum_{k=1}^{m} (-1)^{k-1} \left[\frac{B_k(1)}{k!} f^{(k-1)}(j) - \frac{B_k(0)}{k!} f^{(k-1)}(j-1) \right] \tag{18}$$

$$+ (-1)^m \int_{j-1}^{j} \frac{B_m(x-j+1)}{k!} f^{(m)}(x)\,dx.$$

Now we sum from $j = 1$ to n. This gives

$$\int_{0}^{n} f(x)\,dx = X + (-1)^m \sum_{j=1}^{n} \int_{j-1}^{j} \frac{B_m(x-j+1)}{k!} f^{(m)}(x)\,dx, \tag{19}$$

where X is given by

$$X = \sum_{j=1}^{n} \sum_{k=1}^{m} (-1)^{k-1} \left[\frac{B_k(1)}{k!} f^{(k-1)}(j) - \frac{B_k(0)}{k!} f^{(k-1)}(j-1) \right] \tag{20}$$

or, changing the order of the sum,

$$X = \sum_{k=1}^{m} (-1)^{k-1} \sum_{j=1}^{n} \left[\frac{B_k(1)}{k!} f^{(k-1)}(j) - \frac{B_k(0)}{k!} f^{(k-1)}(j-1) \right]. \tag{21}$$

We rewrite this as

$$X = \sum_{k=1}^{m} (-1)^{k-1} \sum_{j=1}^{n} \frac{B_k(1) - B_k(0)}{k!} f^{(k-1)}(j) \tag{22}$$

$$+ \sum_{k=1}^{m} (-1)^{k-1} \sum_{j=1}^{n} \frac{B_k(0)}{k!} [f^{(k-1)}(j) - f^{(k-1)}(j-1)].$$

Equation (6) makes it easy to evaluate the first sum. In the second we write B_k for $B_k(0)$ and move the factor of $B_k/k!$ outside the inner sum. This gives

$$X = \sum_{j=1}^{n} f(j) + \sum_{k=1}^{m} (-1)^{k-1} \frac{B_k}{k!} \sum_{j=1}^{n} [f^{(k-1)}(j) - f^{(k-1)}(j-1)]. \tag{23}$$

After summing over j all but the terms $f^{(k-1)}(n)$ and $-f^{(k-1)}(0)$ cancel, and we are left with

$$X = \sum_{j=1}^{n} f(j) + \sum_{k=1}^{m} (-1)^{k-1} \frac{B_k}{k!} [f^{(k-1)}(n) - f^{(k-1)}(0)]. \tag{24}$$

Inserting this back in (19) we find

$$\int_0^n f(x)\,dx \;=\; \sum_{j=1}^n f(j) + \sum_{k=1}^m (-1)^{k-1}\frac{B_k}{k!}[f^{(k-1)}(n) - f^{(k-1)}(0)] \quad (25)$$

$$+ (-1)^m \sum_{j=1}^n \int_{j-1}^j \frac{B_m(x-j+1)}{k!} f^{(m)}(x)\,dx.$$

Bringing the second and third terms on the right over to the left gives (13), so the proof of the theorem is complete.

We stated at the beginning of this section that Theorem 5 often allows us to replace the limit of a slowly convergent sequence by that of a rapidly convergent sequence. Suppose, for example, that we wish to calculate

$$\sum_{n=1}^{\infty} n^{-3/2}$$

to eighteen decimal places. This is just $\zeta(3/2)$ where $\zeta(s)$ is defined by

$$\zeta(s) = \sum_{n=1}^{\infty} n^{-s} = \lim_{a\to\infty} \sum_{n=1}^a n^{-s} \quad (26)$$

for $s > 1$. We will have more to say about this function in Section 1.8. By the definition of the infinite sum, we are assured that there is a number a for which

$$|\zeta(\tfrac{3}{2}) - \sum_{n=1}^a n^{-3/2}| \le 5 \cdot 10^{-19}, \quad (27)$$

so we need only compute a sum of a terms. In fact

$$a \approx 1.6 \cdot 10^{37}, \quad (28)$$

so this may take a little while.

A more intelligent strategy, and the one actually employed by Euler to compute the table of values of ζ in [26], is to use the Euler-MacLaurin formula. We apply Theorem 5 to $f(x) = (x+a)^{-s}$ and $n = b - a$, where a and b are positive integers with $a < b$. Equation (13) shows that

$$\sum_{j=1}^{b-a} (j+a)^{-s} = \sum_{k=0}^m \frac{B_k}{k!}(s)_{k-1}[a^{-s-k+1} - b^{-s-k+1}] + e_{a,b,m}(s), \quad (29)$$

or, making the change of variable $n = j + a$,

$$\sum_{n=a+1}^b n^{-s} = \sum_{k=0}^m \frac{B_k}{k!}(s)_{k-1}[a^{-s-k+1} - b^{-s-k+1}] + e_{a,b,m}(s). \quad (30)$$

Note that the integral in equation (13) has not disappeared; it has simply become the $k = 0$ term in the sum on the right. Equation (14) shows that the error term $e_{a,b,m}(s)$ is bounded by

$$|e_{a,b,m}(s)| \leq \frac{B_m}{m!}(s)_{m-1}[a^{-s-m+1} - b^{-s-m+1}] < \frac{B_m}{m!}(s)_{m-1}a^{-s-m+1}. \quad (31)$$

Now we take the limit as b tends to infinity,

$$\zeta(s) - \sum_{n=1}^{a} n^{-s} = \sum_{k=0}^{m} \frac{B_k}{k!}(s)_{k-1}a^{-s-k+1} + e_{a,m}(s), \quad (32)$$

where

$$|e_{a,m}(s)| < \frac{B_m}{m!}(s)_{m-1}a^{-s-m+1}. \quad (33)$$

We can rewrite equation (32) as

$$\zeta(s) = \sum_{n=1}^{a} n^{-s} + \sum_{k=0}^{m} \frac{B_k}{k!}(s)_{k-1}a^{-s-k+1} + e_{a,m}(s). \quad (34)$$

Obviously

$$\lim_{a \to \infty} e_{a,m}(s) = 0 \quad (35)$$

for any $m \geq 0$ and $s > 1$, so we can evaluate $\zeta(s)$ using the limit

$$\zeta(s) = \lim_{a \to \infty} \left\{ \sum_{n=1}^{a} n^{-s} + \sum_{k=0}^{m} \frac{B_k}{k!}(s)_{k-1}a^{-s-k+1} \right\} \quad (36)$$

for any value of m.

To see how effective this strategy is in practice we return to the problem of computing $\zeta(3/2)$ to eighteen decimal places. Taking $a = m = 16$ we see that

$$|e_{a,m}(s)| < 4.44 \cdot 10^{-19}, \quad (37)$$

so the error is within the allowed range. An easy calculation gives

$$\sum_{n=1}^{16} n^{-3/2} + \sum_{k=0}^{16} \frac{B_k}{k!}(3/2)_{k-1}a^{-k-1/2} = 2.612375348685439782881 \quad (38)$$

to twenty-one places. To get this we needed to sum thirty-three terms, of which only twenty-six were nonzero. This certainly compares favorably with $1.6 \cdot 10^{37}$.

In the next section we will apply a variant of the same idea to $s < 1$. There the original series (26) is divergent and the use of the Euler-MacLaurin formula is to find out what the correct definition of $\zeta(s)$ is.

Exercises

1. Compute $\sum_{n=1}^{\infty} n^{-2}$ to as many places as your patience allows. We will see in Section 2.7 that

$$\sum_{n=1}^{\infty} n^{-2} = \frac{\pi^2}{6}.$$

   This formula should help you check your work.

2. Show that equation (16) of Section 1.3 can be improved to

$$1 + \frac{1}{2} + \frac{1}{3} + \cdots + \frac{1}{n} - \log n = \gamma + O(n^{-1}).$$

3. Evaluate γ to many decimal places.

4. In [56] Olivier claimed that if a_n is a sequence for which

$$\lim_{n\to\infty} n a_n = 0$$

   converges, then

$$\sum_{n=1}^{\infty} a_n$$

   converges. Abel disputed this, claiming, without proof, that

$$a_n = \frac{1}{n \log n}$$

   is a counterexample. Who was right?

1.8 The Zeta Function

The series[11]

$$\zeta(s) = \sum_{n=1}^{\infty} n^{-s} \tag{1}$$

is convergent for $s > 1$. In fact comparison with the corresponding integral gives not only convergence, but also the bounds

$$\frac{1}{s-1} \int_{x=1}^{\infty} x^{-s}\, dx \le \sum_{n=1}^{\infty} n^{-s} \le 1 + \int_{x=1}^{\infty} x^{-s}\, dx = \frac{s}{s-1}. \tag{2}$$

---

[11]This function is conventionally known as the Riemann zeta function, although it first appears in the paper [24], written by Euler 89 years before Riemann's birth.

From these we see that

$$\lim_{s \to 1^+} \zeta(s) = \infty \tag{3}$$

while

$$\lim_{s \to 1^+} (s - 1)\zeta(s) = 1, \tag{4}$$

Euler already considered what meaning the sum might be given for $s < 1$. He also determined the values when s is a positive even integer. As we shall see in Section 2.7, these latter are certain rational multiples of π^s.

Writing each positive integer n as the product of its prime factors gives the Euler product formula

$$\zeta(s) = \prod_p (1 - p^{-s})^{-1}, \tag{5}$$

where p runs over primes, here and throughout this section. In more detail the argument goes as follows.

$$\begin{aligned}
\prod_p (1 - p^{-s})^{-1} &= \lim_{P \to \infty} \prod_{p \le P} (1 - p^{-s})^{-1} \\
&= \lim_{P \to \infty} \prod_{p \le P} \lim_{K \to \infty} \sum_{k \le K} p^{-ks} \\
&= \lim_{P \to \infty} \lim_{K \to \infty} \sum n^{-s}.
\end{aligned} \tag{6}$$

The final sum is over all n with no prime divisor greater than P or with multiplicity greater than K. Any n satisfies this requirement for P and K sufficiently large. Let $N(P, K) + 1$ be the smallest number not included in this sum. Since the sum contains positive terms only

$$\sum_{n=1}^{N(P,K)} n^{-s} \le \sum n^{-s} \le \sum_{n=1}^{\infty} n^{-s}, \tag{7}$$

where the middle sum has the same range as before. For any positive N there is some P, K such that $N(P, K) \ge N$, so

$$\sum_{n=1}^{N} n^{-s} \le \prod_p (1 - p^{-s})^{-1} \le \sum_{n=1}^{\infty} n^{-s}. \tag{8}$$

Since N can be chosen arbitrarily large, we see that (5) must hold.

Equations (3) and (5) furnish a proof that there are infinitely many primes. Otherwise we would have

$$\lim_{s \to 1+} = \prod_p (1 - p^{-1}) < \infty.$$

It is interesting to try to make this proof into a quantitative lower bound for the number $\pi(N)$ of primes less N. It seems to be impossible by this method to get a lower bound which grows faster than $\log \log N$, which is the bound one gets from Euclid's purely algebraic argument. More realistic bounds for the number of primes require more information about $\zeta(s)$ than the crude estimates (2).

Our first task is to extend the function $\zeta(s)$ to $s < 1$. In view of equations (3) and (4), we cannot hope to find a smooth extension of $\zeta(s)$. We can, however, look for a smooth extension of $(s - 1)\zeta(s)$. There are several ways to do this. Two are due to Euler. One, which gives an extension only to $s > 0$, is described in the exercises. The other, which gives an extension to all s, is described below. It relies on the Euler-MacLaurin formula. As mentioned in the preceding section, the Euler-MacLaurin formula did not, at that time, have any rigorous foundation and Euler's argument had little influence on the subsequent development of the subject. Two other methods, both of which give an extension to all s, are due to Dirichlet [17] and Riemann [60]. The former is also sketched in the exercises. All of these extensions agree on their common domain of definition. This is an immediate consequence of the rigidity principle, Theorem 18 of Section 2.4, but it is by no means obvious without that theorem.

We begin by defining

$$h_N(x, \rho) = \rho \sum_{n=0}^{N-1} (x + n)^{-\rho - 1} \tag{9}$$

for $x > 0$. For $\rho > 0$ we define

$$H(x, \rho) = \lim_{N \to \infty} h_N(x, \rho) = \rho \sum_{n=0}^{\infty} (x + n)^{-\rho - 1}. \tag{10}$$

The limit on the right is divergent for $\rho < 0$. We will assign $H(x, \rho)$ a value for these values of ρ later. Writing $\rho + 1$ in place of s, the standard convention until Riemann's paper, will make some of the formulae which follow slightly nicer. The introduction of the extra parameter x, an innovation due to Hurwitz, is of greater importance. We will see shortly why it is helpful. We note that $H(x, 1)$ is just the second derivative of $\log \Gamma(x)$. We will examine the connection between H and the Gamma function in more detail at the end of this section. The connection between H and ζ is

$$\zeta(s) = (s - 1)^{-1} H(1, s - 1). \tag{11}$$

We apply Theorem 5 to (9). This gives

$$h_N(x, \rho) = g_{m,N}(x, \rho) - \sum_{k=0}^{m-1} (-1)^k \frac{B_k}{k!} (\rho)_k (x + N)^{-\rho-k} \qquad (12)$$

where

$$\begin{aligned} g_{m,N}(x, \rho) &= \sum_{k=0}^{m-1} (-1)^k \frac{B_k}{k!} (\rho)_k x^{-\rho-k} \qquad (13) \\ &\quad - \sum_{n=0}^{N-1} \int_0^1 \frac{B_m(u)}{m!} (\rho)_{m+1} (x + u + n)^{-\rho-m-1} \, du. \end{aligned}$$

In analogy with (10) we define

$$G_m(x, \rho) = \lim_{N \to \infty} g_{m,N}(x, \rho) = \rho \sum_{n=0}^{\infty} (x + n)^{-\rho-1}. \qquad (14)$$

The reason for introducing the extra parameter x now becomes clear. We can write $g_{m,N}$ in the form

$$\begin{aligned} g_{m,N}(x, \rho) &= \sum_{k=0}^{m-1} (-1)^k \frac{B_k}{k!} (\rho)_k x^{-\rho-k} \qquad (15) \\ &\quad - \int_0^1 \frac{B_m(u)}{m!} (\rho)_{m+1} h_N(x + u, \rho + m + 1) \, du. \end{aligned}$$

$h_N(x + u, \rho + m + 1)$ converges to $H(x + u, \rho + m + 1)$ uniformly in u, provided that $\rho + m + 1 > 0$. For these values of ρ the limit (14) exists and

$$\begin{aligned} G_m(x, \rho) &= \sum_{k=0}^{m-1} (-1)^k \frac{B_k}{k!} (\rho)_k x^{-\rho-k} \qquad (16) \\ &\quad - \int_0^1 \frac{B_m(u)}{m!} (\rho)_{m+1} H(x + u, \rho + m + 1) \, du. \end{aligned}$$

Suppose $\rho + m' + 1 > 0$ and $\rho + m'' + 1 > 0$ and $m' \leq m''$. Then, using equation (12),

$$g_{m'',N}(x, \rho) = g_{m',N}(x, \rho) + \sum_{k=m'}^{m''-1} (-1)^k \frac{B_k}{k!} (\rho)_k (x + N)^{-\rho-k} \qquad (17)$$

and, letting N tend to infinity,

$$G_{m''}(x, \rho) = G_{m'}(x, \rho). \tag{18}$$

A similar argument shows that

$$G_m(x, \rho) = H(x, \rho) \tag{19}$$

if $\rho > 0$. In fact one can regard this as simply the case $m'' = m$, $m' = 0$ of equation (18). Thus we see that the function $G_m(x, \rho)$ extends $H(x, \rho)$ to $\rho + m + 1 > 0$. For any value of ρ there is such an integer m. By equation (18) all such values of m give the same $G_m(x, \rho)$. We may therefore use equation (19) to *define* $H(x, \rho)$ for all $\rho \leq 0$.

Our next task is to check that $H(x, \rho)$ is smooth as a function of ρ. It is also smooth as a function of x, but we do not care. We begin by computing derivatives of h_N,

$$\frac{\partial^j}{\partial \rho^j} h_N(x, \rho) = (-1)^j \rho \sum_{n=0}^{N-1} (x + n)^{-\rho-1} \log^j (x + n) \tag{20}$$

$$+ (-1)^{j-1} j \sum_{n=0}^{N-1} (x + n)^{-\rho-1} \log^{j-1}(x + n).$$

This is a smooth function of ρ. For $\rho > 0$ it converges locally uniformly to

$$\frac{\partial^j}{\partial \rho^j} H(x, \rho) = (-1)^j \rho \sum_{n=0}^{\infty} (x + n)^{-\rho-1} \log^j (x + n) \tag{21}$$

$$+ (-1)^{j-1} j \sum_{n=0}^{\infty} (x + n)^{-\rho-1} \log^{j-1}(x + n).$$

To prove this we use the estimate (23) of the appendix,

$$\log^j (x + n) < \frac{2j}{\rho}(x + n)^{\rho/2} \tag{22}$$

$$\log^{j-1}(x + n) < \frac{2j - 2}{\rho}(x + n)^{\rho/2}.$$

The convergence of the sum (22) follows by the comparison test from that of $H(x, \rho/2)$. In this way we see that $H(x, \rho)$ is a smooth function of ρ. In fact we obtain in this way the quantitative estimate

$$|\frac{\partial^j}{\partial \rho^j} H(x, \rho)| \leq \frac{2j^2}{\rho} H(x, \rho/2), \tag{23}$$

which we will need shortly. Formal differentiation under the integral in (16) gives the formula

$$\frac{\partial^l}{\partial\rho^l}G_m(x,\rho) = \sum_{k=0}^{m-1}\sum_{j=0}^{l}(-1)^k\frac{B_k}{k!}\binom{l}{j}\frac{\partial^{l-j}}{\partial\rho^{l-j}}(\rho)_k x^{-\rho-k}\log^j x \qquad (24)$$

$$-\sum_{j=0}^{l}\binom{l}{j}\frac{\partial^{l-j}}{\partial\rho^{l-j}}(\rho)_{m+1}I_{j,m}(x)$$

where $I_{j,m,x}$ is the integral

$$I_{j,m}(x) = \frac{1}{m!}\int_0^1 B_m(u)\frac{\partial^j}{\partial\rho^j}H(x+u,\rho+m+1)\,du. \qquad (25)$$

The partial derivatives of $(\rho)_k$ are just polynomials in ρ and cause no trouble. The partial derivatives of $H(x+u,\rho+m+1)$ were estimated in (23). This gives the bound

$$|\frac{\partial^j}{\partial\rho^j}H(x+u,\rho+m+1)| \le \frac{2j^2}{\rho+m+1}H(x+u,\rho/2). \qquad (26)$$

Since H is a decreasing function of its first argument, we obtain the uniform bound

$$|\frac{\partial^j}{\partial\rho^j}H(x+u,\rho+m+1)| \le \frac{2j^2}{\rho+m+1}H(x,\rho/2) \qquad (27)$$

on the domain of integration. Differentiation under the integral is justified by Theorem S of the appendix, and we see that $G_m(x,\rho)$ is smooth for $\rho+m+1 > 0$. In view of the definition of $H(x,\rho)$ it follows that H is smooth for all ρ.

From equation (4) we see that $H(1,1) = 1$, so

$$\frac{H(1,\rho)-1}{\rho} \qquad (28)$$

extends to a function smooth for all ρ. From the preceding paragraph we see that

$$\frac{\frac{\partial}{\partial\rho}H(1,\rho)}{H(1,\rho)} \qquad (29)$$

is smooth for all ρ for which $H(1,\rho) \ne 0$. In particular it is smooth for ρ near 0. Thus

$$\frac{H(1,\rho)-1}{\rho} + \frac{\frac{\partial}{\partial\rho}H(1,\rho)}{H(1,\rho)} \qquad (30)$$

extends to a function which is smooth in a neighborhood of 0. In terms of ζ this says that

$$\zeta(s) + \frac{\zeta'(s)}{\zeta(s)} \tag{31}$$

extends to a function which is smooth in a neighborhood of 1. In particular all of its derivatives approach finite limits as s tends to 1 from the right.

For applications to the analytic theory of numbers, the most important part of Euler's investigations concerns the successive derivatives of ζ and its logarithm. Equation (21) above shows that

$$\frac{d^m}{ds^m}\zeta(s) = (-1)^m \sum_{n=1}^{\infty} n^{-s} \log^m n. \tag{32}$$

Formal differentiation of the logarithm of the Euler product formula (5) produces

$$\frac{d^m}{ds^m} \frac{\zeta'(s)}{\zeta(s)} = (-1)^{m+1} \sum_p p^{-s} \log^{m+1} p + e_m(s), \tag{33}$$

where

$$e_m(s) = \sum_{k=2}^{m+2} a_{m,k} \sum_p \frac{p^{-ks} \log^{m+1} p}{(1 - p^{-s})^{m+2}}. \tag{34}$$

The $a_{m,k}$ are certain integers which can be calculated inductively. Thus we see that

$$\sum_{n=1}^{\infty} n^{-s} \log^m n - \sum_p p^{-s} \log^{m+1} p + (-1)^m e_m(s) \tag{35}$$

approaches a finite limit as s tends to 1 from the right. The justification of the term by term differentiation is straightforward. We need to check that the series on the right converge locally uniformly on $s > 1$. In fact the term $e_m(s)$ converges locally uniformly for $s > \frac{1}{2}$, and thus represents a continuous function there. It follows that $e_m(s)$ approaches a finite limit as s tends to 1 from the right, as therefore does

$$\sum_{n=1}^{\infty} n^{-s} \log^m n - \sum_p p^{-s} \log^{m+1} p. \tag{36}$$

In the next section we will see how Chebyshev used this fact to obtain information about the distribution of primes.

The Hurwitz function $H(x, \rho)$ is important not only for the study of the zeta function but also for the study of the Gamma function. We can use Theorem 5 to prove that Stirling's series gives an asymptotic expansion for the Gamma function. We start from Euler's formula

$$\Gamma(z) = \lim_{n\to\infty} \frac{n^z n!}{(z)_{n+1}}. \tag{37}$$

Writing $n!$ as $n\Gamma(n)$ and using the estimates (3) of Section 1.3 we see that

$$\Gamma(z) = \sqrt{2\pi} \lim_{n\to\infty} \frac{n^{z+n+\frac{1}{2}} e^{-n}}{(z)_{n+1}}. \tag{38}$$

Taking logarithms we find

$$\log \Gamma(z) = \frac{1}{2}\log(2\pi) + \lim_{n\to\infty} \{(z+n+\frac{1}{2})\log n - n - \sum_{j=0}^{n}\log(z+j)\}. \tag{39}$$

We apply Theorem 5 to this sum.

$$\begin{aligned}
\sum_{j=1}^{n}\log(z+j) &= (z+n+\frac{1}{2})\log(z+n) - (z+\frac{1}{2})\log z - n \tag{40}\\[2mm]
&\quad + \sum_{k=2}^{m}\frac{B_k}{k(k-1)}z^{1-k} - \sum_{k=2}^{m}\frac{B_k}{k(k-1)}(z+n)^{1-k}\\[2mm]
&\quad + \sum_{j=1}^{n}\int_{j-1}^{j}\frac{B_m(x-j+1)}{m}(z+x)^{-m}\,dx.
\end{aligned}$$

Inserting this in equation (39) gives

$$\begin{aligned}
\log \Gamma(z) &= \frac{1}{2}\log(2\pi) + (z-\frac{1}{2})\log z + \sum_{k=2}^{m}\frac{B_k}{k(k-1)}z^{1-k} \tag{41}\\[2mm]
&\quad + \lim_{n\to\infty}(z+n+\frac{1}{2})\log(\frac{n}{z+n})\\[2mm]
&\quad - \lim_{n\to\infty}\sum_{k=2}^{m}\frac{B_k}{k(k-1)}(z+n)^{1-k}\\[2mm]
&\quad + \lim_{n\to\infty}\sum_{j=1}^{n}\int_{j-1}^{j}\frac{B_m(x-j+1)}{m}(z+x)^{-m}\,dx.
\end{aligned}$$

We now evaluate the limits on the right. By Taylor's theorem

$$\log(\frac{n}{z+n}) = -\log(1+\frac{z}{n}) = -\frac{z}{n} = O(n^{-2}), \tag{42}$$

so

$$\lim_{n\to\infty} (z + n + \frac{1}{2}) \log(\frac{n}{z+n}) = -z. \tag{43}$$

The second limit on the right in (41) is a sum of finitely many terms each of which individually tends to zero, so

$$\lim_{n\to\infty} \sum_{k=2}^{m} \frac{B_k}{k(k-1)}(z+n)^{1-k} = 0. \tag{44}$$

The third limit is just the definition of the infinite sum

$$\sum_{j=1}^{\infty} \int_{j-1}^{j} \frac{B_m(x - j + 1)}{m}(z+x)^{-m} \, dx. \tag{45}$$

We therefore conclude that

$$\log \Gamma(z) = \frac{1}{2} \log(2\pi) + (z - \frac{1}{2}) \log z - z + J(z) \tag{46}$$

where

$$J(z) = \sum_{k=2}^{m} \frac{B_k}{k(k-1)} z^{1-k} + e_m(z). \tag{47}$$

The error term $e_m(z)$ is given by

$$e_m(z) = \sum_{j=1}^{\infty} \int_{j-1}^{j} \frac{B_m(x - j + 1)}{m}(z+x)^{-m} \, dx \tag{48}$$

$$= \sum_{n=0}^{\infty} \int_{0}^{1} \frac{B_m(u)}{m}(z+u+n)^{-m} \, du$$

$$= \int_{0}^{1} \frac{B_m(u)}{m(m-1)} H(z+u, m-1) \, du.$$

and hence satisfies the estimate

$$|e_m(z)| \leq \max_{0 \leq u \leq 1} \frac{B_m(u)}{m(m-1)} \int_{0}^{1} |H(z+u, m-1)| \, du. \tag{49}$$

We estimate $\int_{0}^{1} |H(z+u, m-1)| \, du$ as follows.

$$|H(z+u, m-1)| \leq \rho \sum_{n=0}^{\infty} |z+u+n|^{-\rho-1}, \tag{50}$$

so

$$\int_0^1 |H(z+u, m-1)| \, du \le \rho \sum_{n=0}^\infty \int_0^1 |z+u+n|^{-\rho-1} \, du. \qquad (51)$$

Making the substitution $v = u + n$, we find

$$\int_0^1 |H(z+u, m-1)| \, du \le \rho \sum_{n=0}^\infty \int_n^{n+1} |z+v|^{-\rho-1} \, dv = \rho \int_0^\infty |z+v|^{-\rho-1} \, dv.$$

$$(52)$$

Since

$$\rho \int_0^\infty |z+v|^{-\rho-1} \, dv = z^{-\rho} \qquad (53)$$

we obtain the estimate

$$|e_m(z)| \le \max_{0 \le u \le 1} \frac{B_m(u)}{m(m-1)} z^{1-m} \qquad (54)$$

It follows that

$$\Gamma(z) = \sqrt{\frac{2\pi}{z}} (\frac{z}{e})^z e^{J(z)} \qquad (55)$$

where $J(z)$ has the asymptotic series

$$\sum_{k=2}^\infty \frac{B_k}{k(k-1)} z^{1-k}. \qquad (56)$$

The preceding proof is by no means the most concise possible. It has the advantage, however, that it requires almost no modification for complex z. In fact everything up to equation (53) can be used without change. This will be important in Section 2.9.

Exercises

1. This exercise outlines Euler's method for extending ζ to $(0, \infty)$. First prove the identity

$$(1 - 2^{1-s})\zeta(s) = L(s)$$

for $s > 1$, where

$$L(s) = \sum_{n-1}^\infty (-1)^{n+1} n^{-s}.$$

Then show, by summation by parts, that the sum defining $L(s)$ converges locally uniformly for $s > 0$.

2. This exercise and the one which follows are based on Dirichlet's paper [17]. Suppose $s > 1$. Prove the integral representation

$$\Gamma(s)\zeta(s) \doteq \int_0^\infty \frac{t^{s-1}\,dt}{e^t - 1}.$$

Formally this can be obtained from the series expansion

$$\frac{1}{e^t - 1} = \sum_{n=1}^\infty e^{-nt}$$

by integrating term by term against $t^{s-1}\,dt$. Justify the interchange of the limit and integral. You may find it easiest to work with the finite partial sums

$$\sum_{n=1}^N e^{-nt} = \frac{1}{e^t - 1} - \frac{e^{-Nt}}{e^t - 1}$$

and delay the limit as N tends to infinity until the last possible moment.

3. The main point of the integral representation of the preceding exercise is that it can be evaluated in a completely different way. Chebyshev, in his paper [9], seems to have been the first to exploit this idea. Chebyshev's idea was to rewrite the integral in the form

$$\Gamma(s)\zeta(s) = \int_0^\infty \left(\frac{te^t}{e^t - 1}\right) t^{s-2} e^{-t}\,dt$$

and use the expansion (4) of Section 1.7. Formal term by term integration gives the divergent series

$$\zeta(s) \stackrel{?}{=} \frac{1}{s-1} + \sum_{k=1}^\infty \frac{B_k(1)}{k!}(s)_{k-1}.$$

Nothing we can do will make this divergent series converge, but it can be used to extend the definition of $\zeta(s)$. Replacing the infinite sum (4) by Taylor's series with remainder, and then integrating term by term gives a series of the form

$$\Gamma(s)\zeta(s) = \sum_{k=0}^{m-1} \frac{B_k(1)\Gamma(s+k-1)}{k!} r_m(s)\,dt.$$

Find an explicit expression for $r_m(s)$ and use it to show that $r_m(s)$, and hence $\zeta(s) - \frac{1}{s-1}$ are smooth on $(1 - m, \infty)$.

1.9   The Distribution of Primes II

Armed with the zeta function we can now return to the problem of the distribution of primes. We wish to show that

$$d(n) = \pi(n) - \text{li}\,n \tag{1}$$

is sometimes small. Chebyshev's other main theorem, proved in [9], is

THEOREM 6 (CHEBYSHEV'S FIRST THEOREM)
For any positive δ and any nonnegative integers m and N, the inequality

$$|d(n)| \leq \frac{\delta n}{\log^m n} \tag{2}$$

holds for some $n \geq N$.

In Section 2.5 we will see how some relatively simple ideas from complex analysis can be used to derive a much sharper theorem.

Proof.

$$\pi(n) \leq \pi(n+1) \leq \pi(n) + 1 \tag{3}$$

and

$$\text{li}\,n \leq \text{li}(n+1) \leq \text{li}\,n + 1, \tag{4}$$

so

$$|d(n+1) - d(n)| \leq 1. \tag{5}$$

If $d(n+1)$ and $d(n)$ are of opposite sign then at least one of them is of absolute value at most $\frac{1}{2}$.

Most of the effort below is devoted to estimating "error terms." Though important, this tends to hide the interesting part of the proof, which is the manipulation of the "main term." There is an additional complication. We will revisit this problem later using the techniques of complex analysis. If we wish to avoid much duplication of effort we must make our estimates on the error terms uniform in m as well as s. For computations which are purely routine, even if very long, we will simply give the results. We will consider only s in the interval $(1, 2)$.

Our starting point is the calculation at the end of the preceding section, which showed that

$$\lim_{r \to \infty} Y(m, r, s) \tag{6}$$

converges to a limit which is bounded for $s > 1$, where

$$Y(m, r, s) = \sum_{n=2}^{r} n^{-s} \log^{m-1} n - \sum_{p} p^{-s} \log^m p \qquad (7)$$

is bounded uniformly for $s > 1$, independently of r and s. We write S_m for an upper bound for this sum,

$$|\lim_{r \to \infty} Y(m, r, s)| \leq S_m. \qquad (8)$$

Since

$$\pi\left(n + \frac{1}{2}\right) - \pi\left(n - \frac{1}{2}\right) = \begin{cases} 1 & n \text{ prime} \\ 0 & n \text{ composite} \end{cases} \qquad (9)$$

we can rewrite (7) as

$$Y(m, r, s) = \sum_{n=2}^{r} \left(\pi\left(n + \frac{1}{2}\right) - \pi\left(n - \frac{1}{2}\right) - \frac{1}{\log n}\right) n^{-s} \log^m n. \qquad (10)$$

We approximate $\frac{1}{\log n}$ by $\int_{n-\frac{1}{2}}^{n+\frac{1}{2}} \frac{du}{\log u}$ using inequality (81) from the appendix,

$$\left| \frac{1}{\log n} - \int_{n-\frac{1}{2}}^{n+\frac{1}{2}} \frac{du}{\log u} \right| \leq \frac{1}{24} \max_{n-\frac{1}{2} \leq u \leq n+\frac{1}{2}} \left| \frac{d^2}{du^2} \frac{1}{\log u} \right| \qquad (11)$$

$$\leq \frac{1}{24(n - \frac{1}{2})^2} \left[\frac{1}{\log^2(n - \frac{1}{2})} + \frac{2}{\log^3(n - \frac{1}{2})} \right]$$

$$\leq C_1 n^{-2}$$

with a constant

$$C_1 = \frac{2}{9} \log^3\left(\frac{3}{2}\right) \qquad (12)$$

independent of m, n, r, and s. It follows that

$$Y(m, r, s) = \sum_{n=2}^{r} \left(d\left(n + \frac{1}{2}\right) - d\left(n - \frac{1}{2}\right)\right) n^{-s} \log^m n + E_1(m, r, s), \qquad (13)$$

with an error term

$$|E_1(m, r, s)| \leq C_1 \sum_{n=2}^{\infty} n^{-s-2} \log^m n. \qquad (14)$$

This sum can be put in a more convenient form using summation by parts. We write (13) as

$$Y(m, r, s) = \sum_{n=2}^{r} (a_n - a_{n-1}) b_n + E_1(m, r, s), \qquad (15)$$

where

$$a_n = d(n + \frac{1}{2}) + \text{li} \frac{3}{2} \qquad (16)$$

$$b_n = n^{-s} \log^m n.$$

Since $a_1 = 0$ it follows from (5) that

$$|a_n| \leq n.$$

and hence

$$|a_r b_r| \leq r^{1-s} \log^m r. \qquad (17)$$

We will write $\epsilon(m, r, s)$ for $a_r b_r$. We conclude that the sum (15) can be rewritten as

$$Y(m, r, s) = \sum_{n=1}^{r-1} a_n (b_{n+1} - b_n) + E_1(m, r, s) + \epsilon(m, r, s), \qquad (18)$$

where

$$\epsilon(m, r, s) \leq r^{1-s} \log^m r. \qquad (19)$$

Next we approximate the difference

$$b_{n+1} - b_n = (n + 1)^{-s} \log^m (n + 1) - n^{-s} \log^m n \qquad (20)$$

by the derivative

$$\frac{d}{dx} \{x^{-s} \log^m x\}|_{x=n} = (m - s \log n) n^{-s-1} \log^{m-1} n. \qquad (21)$$

using inequality (85) of the appendix. This gives

$$|b_{n+1} - b_n - \frac{d}{dx} \{x^{-s} \log^m x\}|_{x=n}| \leq \frac{1}{2} \max_{n \leq x \leq n+1} |\frac{d^2}{dx^2} \{x^{-s} \log^m x\}|. \qquad (22)$$

A (rather bad) estimate on the derivative is

$$\max_{n \leq x \leq n+1} |\frac{d^2}{dx^2}\{x^{-s} \log^m x\}| \leq 8(m+1)(m+2)n^{-s-2} \log^m n. \tag{23}$$

Thus

$$|b_{n+1} - b_n - \frac{d}{dx}\{x^{-s} \log^m x\}|_{x=n}| \leq C_2(m+1)(m+2)n^{-s-2} \log^m n \tag{24}$$

with a constant $C_2 = 4$ independent of m, n, r, and s. We conclude that

$$Y(m, r, s) = \sum_{n=1}^{r-1} a_n(m - s \log n)n^{-s} \log^{m-1} n \tag{25}$$
$$+ E_1(m, r, s) + E_2(m, r, s) + \epsilon(m, r, s)$$

with

$$|E_2(m, r, s)| \leq C_2(m+1)^2 \sum_{n=2}^{\infty} n^{-s-1} \log^m n. \tag{26}$$

There are no integers, and thus no primes, in the interval $(n, n + \frac{1}{2}]$, so

$$\pi(n + \frac{1}{2}) = \pi(n). \tag{27}$$

Estimating the integral by the maximum of the absolute value, we find

$$| \operatorname{li}(n + \frac{1}{2}) - \operatorname{li} n| \leq \frac{1}{2 \log n} \leq \frac{1}{2 \log 2}. \tag{28}$$

From these two equations we see that

$$|a_n - d(n)| \leq C_3, \tag{29}$$

where

$$C_3 = \frac{1}{2 \log 2} + \operatorname{li} \frac{3}{2} \tag{30}$$

is independent of m, n, r, and s. Thus

$$Y(m, r, s) = \sum_{n=1}^{\infty} d(n)(m - s \log n)n^{-s-1} \log^{m-1} n \tag{31}$$
$$+ E_1(m, r, s) + E_2(m, r, s) + E_3(m, r, s)$$
$$+ \epsilon(m, r, s),$$

where

$$|E_3(m, r, s)| \le C_3(m + 1) \sum_{n=2}^{\infty} n^{-s-1} \log^m n. \tag{32}$$

We now combine the error terms $E_j(m, r, s)$ into one,

$$Y(m, r, s) = \sum_{n=1}^{\infty} d(n)(m - s \log n)n^{-s-1} \log^{m-1} n \tag{33}$$
$$+ E(m, r, s) + \epsilon(m, r, s),$$

where

$$|E(m, r, s)| \le C(m + 1)(m + 2) \sum_{n=2}^{\infty} n^{-s-1} \log^m n. \tag{34}$$

with a constant

$$C = C_1 + C_2 + C_3 \tag{35}$$

independent of m, n, r, and s. This gives

$$Y(m, r, s) = mX(m - 1, r - 1, s) - sX(m, r, s) + E(m, r, s), \tag{36}$$

where

$$X(m, r, s) = \sum_{n=2}^{r} d(n)n^{-s-1} \log^m n. \tag{37}$$

Expressing X in terms of Y, we find

$$\frac{X(m, r, s)s^{m+1}}{m!} = \sum_{j=1}^{m} \frac{E(j, s) - Y(j, s)}{j!} s^j + X(0, s)s, \tag{38}$$

a formula which we will need in Section 2.5.

Let m, N, and δ be given. Choose an $N' > N$ such that

$$\delta \frac{n + 1}{\log^m(n + 1)} + \delta \frac{n}{\log^m n} > 1 \tag{39}$$

and

$$\log n > m + 1 \tag{40}$$

for all $n \geq N'$. Define

$$S = S_m + Z + 2 + 2C(m+1)(m+2) \sum_{n=2}^{\infty} n^{-2} \log^m n \qquad (41)$$

where

$$Z = \max_{1 \leq s \leq 2} \left| \sum_{n=2}^{N'-1} d(n)(m - s \log n)n^{-s-1} \log^{m-1} n \right| \qquad (42)$$

Let N'' be sufficiently large that

$$\delta \log N'' > S + \delta \log N' + 1. \qquad (43)$$

Since u^{-1} is a decreasing function

$$\log N'' - \log N' = \int_{N'}^{N''} \frac{du}{u} = \sum_{n=N'}^{N''-1} \int_{n}^{n+1} \frac{du}{u} \leq \sum_{n=N'}^{N''-1} n^{-1}. \qquad (44)$$

It follows that

$$\delta \sum_{N'}^{N''-1} n^{-1} > S + 1. \qquad (45)$$

The function

$$f(\sigma) = \delta \sum_{n=N'}^{N''-1} n^{-\sigma}$$

is continuous at $\sigma = 1$, so, for $s > 1$ and sufficiently close to 1, we have

$$\delta \sum_{N'}^{N''-1} n^{-s} > S. \qquad (46)$$

Choosing such an s, we select an $r > N''$ sufficiently large that

$$\sum_{n=r}^{\infty} (m - s \log n)n^{-s} \log^{m-1} n + n^{1-s} \log^m n < 1 \qquad (47)$$

for all $n \geq r$. We claim that (2) holds for some n in the range $N \leq n \leq r$.
    Because of (39) we know that, if

$$d(n) \leq -\delta n \log^{-m} n \qquad (48)$$

holds for some $n = n_-$ between N' and r and

$$d(n) \geq \delta n \log^{-m} n \qquad (49)$$

holds for some $n = n_+$ between N' and r then there is some n between n_- and n_+ for which (2) holds. There are thus three possibilities:

1. The inequality (2) holds for some $N' \leq n \leq r$, or

2. the inequality (48) holds for all $N' \leq n \leq r$, or

3. the inequality (49) holds for all $N' \leq n \leq r$.

We must show that possibilites 2 and 3 cannot occur. The proofs are almost identical so we will consider only 2. If 2 holds then, using the inequalities (40) and (46),

$$\sum_{n=N'}^{r} d(n)(m - s \log n)n^{-s-1} \log^{m-1} n > S. \tag{50}$$

We estimate the sum from 2 to n' by its maximum,

$$\sum_{n=2}^{r} d(n)(m - s \log n)n^{-s-1} \log^{m-1} n > S - Z. \tag{51}$$

Equation (33) gives

$$Y(m, r, s) \geq S - Z + E(m, r, s) + \epsilon(m, r, s) \tag{52}$$

or, using the estimates (19) and (34) and the definition of S,

$$Y(m, r, s) \geq S_m + 1 \tag{53}$$

This violates (8), so the assumption (48) is untenable.

Chapter 2

Analytic Functions

Now we begin to introduce the basic techniques of complex analysis. It will be some pages yet before they begin to do useful work, but in the end they will prove enormously powerful.

2.1 Contour Integration

In ordinary one-variable calculus classes one learns to integrate any rational function of $\sin\theta$ and $\cos\theta$ over the interval $[0, 2\pi]$. Here we will learn an alternate method which, after a little practice, is easier than the one usually taught. More importantly, it is a simple example of the general technique called the calculus of residues. The calculus of residues, introduced by Cauchy in [7], is perhaps the most powerful technique of complex analysis. Later sections of this chapter will illustrate some of that power.

Using Euler's identities

$$
\begin{aligned}
\sin\theta &= \frac{e^{i\theta} - e^{-i\theta}}{2i}, \\
\cos\theta &= \frac{e^{i\theta} + e^{-i\theta}}{2},
\end{aligned}
\tag{1}
$$

we can write any such integral in the form

$$
\frac{1}{2\pi} \int_0^{2\pi} f(e^{i\theta})\, d\theta,
\tag{2}
$$

where f is a rational function of one variable. Consider now the formal substitution $z = e^{i\theta}$ which transforms the integral into

$$
\frac{1}{2\pi i} \int \frac{f(z)\, dz}{z}.
$$

J. Stalker, *Complex Analysis: Fundamentals of the Classical Theory of Functions*,
Modern Birkhäuser Classics, DOI 10.1007/978-0-8176-4919-7_2,
© Birkhäuser Boston, a part of Springer Science + Business Media, LLC 2009

According to the usual rules of the calculus, one should expect both the upper and lower limits of integration to be 1, and the integral to vanish. But we know, for example that

$$\frac{1}{2\pi} \int_0^{2\pi} 1 \, d\theta = 1 \neq 0, \tag{3}$$

so clearly some care is required in making complex substitutions.

The following definitions cover almost all the cases of interest to us. A *path* is a twice continuously differentiable function γ from a closed interval $[a, b]$ to an open subset $\Omega \in \mathbf{C}$. For example

$$\gamma(\theta) = e^{i\theta}, \quad 0 \leq \theta \leq 2\pi \tag{4}$$

is a path in $\mathbf{C}$. We define the integral along γ of a complex-valued function f, defined and continuous at least on the image of γ, by

$$\int_a^b f(\gamma(t)) \frac{d\gamma}{dt}(t) \, dt. \tag{5}$$

In contrast to the calculus of functions of a real variable, the integral depends on the choice of path, not merely the endpoints. Thus if $\gamma(a) = \gamma(b)$, in which case we say the path is *closed*, it may still happen that the integral is nonzero, as we saw in (3).

The integral is, however, independent of parameterization. In other words, if $\iota: [a, b] \to [c, d]$ is monotone increasing and continuously differentiable along with its inverse, and the path $\tilde{\gamma}$ is defined by

$$\tilde{\gamma}(s) = \gamma(\iota^{-1}(s)),$$

then

$$\int_a^b f(\gamma(t)) \frac{d\gamma}{dt}(t) \, dt = \int_c^d f(\tilde{\gamma}(s)) \frac{d\tilde{\gamma}}{ds}(s) \, ds, \tag{6}$$

as we can see by making the (real) change of variable $s = \iota(t)$ in the integral on the right. It is convenient to assign a special name to classes of paths modulo parameterization. We call these *contours*, and follow the useful notational convention that the contour corresponding to a path, denoted by γ with some accompanying marks, is denoted by C, with the same marks, e.g., that γ_{17} and $\tilde{\gamma}$ belong to the contours C_{17} and $\tilde{C}$ respectively. Since the integral depends only on the contour, we write the integral as

$$\int_C f(z) \, dz. \tag{7}$$

This notation has two advantages. First, we may, whenever convenient, assume that the domain of γ to be any convenient interval, e.g., [0, 1]. Second, we may specify a contour by describing the image of the path and its orientation without bothering to choose a particular parameterization. Thus instead of saying that C is the contour corresponding to the path γ given by equation (4), we can just say that C is the unit circle, with the counterclockwise orientation.

We will need the notion of a *homotopy* between closed paths γ' and γ'' in Ω. We may suppose for convenience that both are parameterized by the interval [0, 1]. A twice continuously differentiable function $\sigma : [0, 1] \times [0, 1] \to \Omega$ is called a homotopy between γ' and γ'' if

$$\sigma(0, t) = \gamma'(t) \tag{8}$$

and

$$\sigma(1, t) = \gamma''(t) \tag{9}$$

for all $0 \le t \le 1$. The corresponding contours are said to be *homotopic*. A contour is said to be *contractible* if it is homotopic to a constant path.

When we estimate integrals we will need the quantity

$$\int_a^b f(\gamma(t)) |\frac{d\gamma}{dt}(t)| \, dt.$$

This is also independent of parameterization and is denoted

$$\int_C f(z) \, |dz|. \tag{10}$$

We will often use the estimate

$$|\int_C f(z) \, dz| \le \int_C |f(z)| \, |dz|, \tag{11}$$

or even the weaker estimate

$$|\int_C f(z) \, dz| \le \max_{z \in C} |f(z)| \int_C |dz|. \tag{12}$$

Of course $z \in C$ means that $z = \gamma(t)$ for some $a \le t \le b$. This notion is independent of the parameterization. The integral on the right is called the *length* of the contour C.

In the ordinary calculus the most useful tool for computing integrals is the fundamental theorem of calculus. This has an analogue for contour integrals, though it is not nearly so important. Suppose

$$f(x) = g'(x) = \lim_{y \to x} \frac{f(y) - f(x)}{y - x}, \tag{13}$$

at least for x in the image of γ. Then

$$
\frac{d}{dt} g(\gamma(t)) = \lim_{s \to t} \frac{g(\gamma(s)) - g(\gamma(t))}{s - t} \tag{14}
$$

$$
= \lim_{s \to t} \frac{g(\gamma(s)) - g(\gamma(t))}{\gamma(s) - \gamma(t)} \lim_{s \to t} \frac{\gamma(s) - \gamma(t)}{s - t}
$$

$$
= f(\gamma(t)) \frac{d\gamma}{dt}(t).
$$

The fundamental theorem of calculus shows that

$$
\int_C f(z)\,dz = \int_a^b f(\gamma(t)) \frac{d\gamma}{dt}(t)\,dt \tag{15}
$$

$$
= \int_a^b \frac{d}{dt} g(\gamma(t))\,dt
$$

$$
= g(\gamma(b)) - g(\gamma(a)).
$$

In particular

$$
\int_C f(z)\,dz = 0 \tag{16}
$$

if the contour is closed.

Suppose now that C is a closed contour and r is a rational function, written as p/q, where p and q are polynomials without common factors. We define, for each $\zeta \in \mathbf{C}$, an integer $v_\zeta(r)$, called the *order* of r at ζ, as follows. If ζ is a zero of p then $v_\zeta(r)$ is its multiplicity, in other words, the number of times $z - \zeta$ divides $p(z)$. If ζ is a zero of q then $v_\zeta(r)$ is minus its multiplicity. Otherwise $v_\zeta(r) = 0$. Those ζ with $v_\zeta(r) < 0$, which are just the zeroes of q, are called the *poles* of r. There are, of course, only finitely many of these. Those poles with $v_\zeta(r) = -1$ are called *simple*. It is traditional to extend r to all of $\mathbf{C}$ by declaring it to be infinite at the poles. It is also traditional to define $r(\infty)$ to be $\lim_{z \to \infty} r(z)$ if it exists and ∞ otherwise. In this way r becomes a function from $\mathbf{C} \cup \infty$ to itself.

Borrowing the theory of partial fractions from algebra, we can find numbers $c_{\zeta,n}$ and $c_{\infty,n}$ such that

$$
r(z) = \sum_{\zeta,n} c_{\zeta,j}(z - \zeta)^{-n} + \sum_{n=0}^{\deg p - \deg q} c_{\infty,n} z^n. \tag{17}
$$

The first sum ranges over all points $\zeta \in C$ and all positive integers n such that $v_\zeta(r) + n < 0$. This is a finite sum. The proof that such an expansion exists

depends crucially on the fundamental theorem of algebra, which we will prove in
Section 2.6.

We note that $(z - \zeta)^{-n}$, for $n > 1$, is the derivative of $(1 - n)^{-1}(z - \zeta)^{1-n}$
while z^n, for $n \geq 0$, is the derivative of $(n + 1)^{-1}z^{n+1}$. We know therefore that
their integrals along C vanish. It follows that

$$\frac{1}{2\pi i} \int_C r(z)\, dz = \sum_\zeta c_{\zeta,1} \frac{1}{2\pi i} \int_C \frac{dz}{z - \zeta}. \tag{18}$$

The number

$$\frac{1}{2\pi i} \int_C \frac{dz}{z - \zeta}$$

is called the *index* of the contour C with respect to the point ζ, and is denoted
by $\mathrm{Ind}(C, z)$. The number $c_{\zeta,1}$ is called the *residue* of r at ζ, and is denoted by
$\mathrm{Res}_{z=\zeta}\, r(z)$. Equation (18) is important enough to be called a theorem.

THEOREM 7 (RESIDUE THEOREM FOR RATIONAL FUNCTIONS)
If C is a closed curve and r a rational function then

$$\frac{1}{2\pi i} \int_C r(z)\, dz = \sum_\zeta \mathrm{Ind}(C, \zeta) \operatorname*{Res}_{z=\zeta} r(z), \tag{19}$$

*where the sum is taken over the poles of r. In words, the contour integral is $2\pi i$
times the sum of the residues, weighted by the indices.*

The point of this theorem is that it organizes the calculation of the integral
on the left hand side into two steps. The first step, the calculation of the indices,
refers only to the contour C; it is independent of the function r. The second step,
the calculation of the residues, refers only to the function r; it is independent of
the contour C.

For poles of low order, especially simple poles, the easiest way to compute
residues is from the formula

$$\operatorname*{Res}_{z=\zeta} r(z) = \frac{1}{(p - 1)!} \lim_{z \to \zeta} \frac{d^{p-1}}{dz^{p-1}} (z - \zeta)^p r(z), \tag{20}$$

where

$$p = -v_\zeta(r) \tag{21}$$

is the order of the pole. This reduces to

$$\operatorname*{Res}_{z=\zeta} r(z) = \lim_{z \to \zeta} (z - \zeta) r(z) \tag{22}$$

for simple poles. These formulae are immediate consequences of (17), the partial fraction expansion.

The indices of C with respect to the poles are usually found "by inspection." One rarely feels the need to give a formal justification, but, when one wishes to do so, the properties in the following theorem are normally sufficient.

THEOREM 8 (PROPERTIES OF THE INDEX)
The index $\text{Ind}(C, \zeta)$ *satisfies the following four conditions.*

1. $\text{Ind}(C, \zeta)$ *is an integer.*

2. $\text{Ind}(C, \zeta)$ *is locally constant on the complement of the image of* γ.

3. *If* γ *and* $\tilde{\gamma}$ *are homotopic paths in* $\mathbf{C} - \{\zeta\}$ *then*

$$\text{Ind}(C, \zeta) = \text{Ind}(\tilde{C}, \zeta). \tag{23}$$

4. *If* γ *is contractible then*

$$\text{Ind}(C, \zeta) = 0. \tag{24}$$

Proof. Choose a path $\gamma : [a, b] \to \mathbf{C} - \{\zeta\}$ belonging to the contour C. Consider the function

$$u(t) = \exp\left(\int_a^t \frac{\frac{d\gamma}{dt}(s)\, ds}{\gamma(t) - \zeta} \right). \tag{25}$$

Clearly

$$u(b) = \exp(2\pi i \, \text{Ind}(C, \zeta)). \tag{26}$$

On the other hand

$$u'(t) = u(t) \frac{\frac{d\gamma}{dt}(t)}{\gamma(t) - \zeta}, \tag{27}$$

so

$$\frac{d}{dt} \frac{u(t)}{\gamma(t) - \zeta} = 0. \tag{28}$$

Since $u(a) = 1$ we must have

$$u(t) = \frac{\gamma(t) - \zeta}{\gamma(a) - \zeta}. \tag{29}$$

Since C is closed, $\gamma(a) = \gamma(b)$ and $u(b) = 1$. In terms of the index this says that

$$\exp(2\pi i \, \text{Ind}(C, \zeta)) = 1. \tag{30}$$

This requires $\text{Ind}(C, \zeta)$ to be an integer, so condition 1 is satisfied.

The integral which defines $\text{Ind}(C, \zeta)$ is continuous in ζ by Theorem R of the appendix. A continuous integer-valued function must be locally constant. This establishes property 2.

Consider the function

$$I(s) = \frac{1}{2\pi i} \int_0^1 \frac{\frac{\partial}{\partial t}\sigma(s, t)\, dt}{\sigma(s, t) - \zeta}. \tag{31}$$

This is a continuous function of s. $I(s)$ is the index of the curve C_s corresponding to the path

$$\gamma_s(t) = \sigma(s, t), \quad 0 \le t \le 1, \tag{32}$$

and hence an integer. A continuous integer-valued function must be locally constant. But $[0, 1]$ is connected, so a locally constant function is necessarily constant, and $I(0) = I(1)$. This proves property 3.

It is clear from the definition that the index of a constant path is zero, so property 4 follows immediately from property 3. This completes the proof of the theorem.

The index of C with respect to ζ captures the intuitive notion of "the number of times C wraps around ζ in the counterclockwise direction." Proving this involves giving this notion a precise definition, more a matter for topology than complex analysis.

As an example we calculate the index of the contour C corresponding to the path γ defined by equation (4). Clearly

$$\text{Ind}(C, 0) = \frac{1}{2\pi i} \int_0^{2\pi} i \, d\theta = 1 \tag{33}$$

and

$$\lim_{\zeta \to \infty} \text{Ind}(C, \zeta) = \lim_{\zeta \to \infty} \frac{1}{2\pi i} \int_C \frac{dz}{z - \zeta} = 0, \tag{34}$$

the interchange of the limit and integral being justified by uniform convergence. Theorem 8 then shows that

$$\text{Ind}(C, \zeta) = \begin{cases} 1 & \text{if } |\zeta| < 1 \\ 0 & \text{if } |\zeta| > 1. \end{cases} \tag{35}$$

It is instructive to give a different proof of (35). If $|\zeta| < 1$ we can expand $\frac{1}{z-\zeta}$ in powers of ζ,

$$\frac{1}{z-\zeta} = \sum_{j=1}^{\infty} \zeta^{j-1} z^{-j} \tag{36}$$

with uniform convergence for $|z| = 1$. We may therefore integrate term by term,

$$\mathrm{Ind}(C, \zeta) = \sum_{j=1}^{\infty} \zeta^{j-1} \frac{1}{2\pi i} \int_C z^{-j} \, dz \tag{37}$$

All terms are zero, except the first, which we have already calculated to be 1. If $|\zeta| > 1$ we can expand $\frac{1}{z-\zeta}$ in powers of ζ^{-1},

$$\frac{1}{z-\zeta} = \sum_{j=0}^{\infty} \zeta^{-j-1} z^{j} \tag{38}$$

again with uniform convergence for $|z| = 1$. Integrating term by term,

$$\mathrm{Ind}(C, \zeta) = \sum_{j=0}^{\infty} \zeta^{-j-1} \frac{1}{2\pi i} \int_C z^{j} \, dz \tag{39}$$

and all terms in the sum are zero.

Exercises

1. Evaluate

$$\frac{1}{2\pi} \int_0^{2\pi} \frac{d\theta}{a \cos \theta + b}$$

   where $|a| < |b|$. Do not assume a and b are real.

2. Let $r = p/q$, with p and q polynomials such that $\deg p < 1 + \deg q$, and let C_ρ be the circle of radius ρ about the origin, oriented counterclockwise. Prove that

$$\lim_{\rho \to \infty} \int_{C_\rho} r(z) \, dz = 0.$$

3. With notation as in the preceding problem, show that

$$\int_{C_\rho} r(z) \, dz = 0$$

   whenever

$$\rho > \max_{q(z)=0} |z|.$$

4. Prove that the order v_ζ is multiplicative in the sense that

$$v_\zeta(fg) = v_\zeta(f) + v_\zeta(g).$$

5. Give an example of a contour C and a point ζ such that $\text{Ind}(\zeta, C) < 0$.

6. Is it possible to have $\text{Ind}(\zeta, C) = 2$?

2.2 Analytic Functions

The limit

$$\lim_{z \to \zeta} \frac{f(z) - f(\zeta)}{z - \zeta} \tag{1}$$

and the contour integral

$$\int_C f(z)\, dz \tag{2}$$

along a closed contour, both of which played a prominent role in the preceding section, are closely related.

Let Ω be an open subset of $\mathbf{C}$. Let $\tilde{\Omega}$ be the open subset of $\mathbf{R}^2$

$$\tilde{\Omega} = \{(x, y) \in \mathbf{R}^2 : x + iy \in \Omega\}. \tag{3}$$

We identify complex-valued functions f on Ω with pairs (u, v) of real-valued functions on $\tilde{\Omega}$ according to the rule

$$f(x + iy) = u(x, y) + iv(x, y), \tag{4}$$

We define functions $\frac{\partial f}{\partial z}$ and $\frac{\partial f}{\partial \bar{z}}$ on Ω by

$$\frac{\partial f}{\partial z}(x + iy) = \left(\frac{1}{2}\frac{\partial}{\partial x} - \frac{i}{2}\frac{\partial}{\partial y}\right)(u(x, y) + iv(x, y)) \tag{5}$$

$$= \frac{1}{2}\left(\frac{\partial u}{\partial x} + \frac{\partial v}{\partial y}\right)(x, y) + \frac{i}{2}\left(\frac{\partial v}{\partial x} - \frac{\partial u}{\partial y}\right)(x, y),$$

$$\frac{\partial f}{\partial \bar{z}}(x + iy) = \left(\frac{1}{2}\frac{\partial}{\partial x} + \frac{i}{2}\frac{\partial}{\partial y}\right)(u(x, y) + iv(x, y)) \tag{6}$$

$$= \frac{1}{2}\left(\frac{\partial u}{\partial x} - \frac{\partial v}{\partial y}\right)(x, y) + \frac{i}{2}\left(\frac{\partial v}{\partial x} + \frac{\partial u}{\partial y}\right)(x, y).$$

With this notation we have the following theorem.

THEOREM 9 (EQUIVALENT DEFINITIONS OF ANALYTICITY)
With notation as above, suppose that u and v are continuously differentiable on $\tilde{\Omega}$. The following conditions are equivalent.

1. $\lim_{\Delta z \to 0} \frac{f(z+\Delta z) - f(z)}{\Delta z}$ *exists for all* $z \in \Omega$.

2. $\frac{\partial f}{\partial \bar{z}} = 0$ *throughout* Ω.

3. *u and v satisfy the* Cauchy-Riemann equations

$$\frac{\partial u}{\partial x} = \frac{\partial v}{\partial y}, \quad \frac{\partial v}{\partial x} = -\frac{\partial u}{\partial y} \tag{7}$$

throughout $\tilde{\Omega}$.

4. $\int_C f(z)\,dz = 0$ *along any closed contractible contour C in* Ω.

5. $\int_{C'} f(z)\,dz = \int_{C''} f(z)\,dz$ *whenever C' and C'' are homotopic.*

Functions satisfying any of these conditions are called *analytic*.

Proof. We begin by showing that conditions 2 and 3 are equivalent. $\frac{\partial f}{\partial \bar{z}}$ is zero if and only if its real and imaginary parts are zero. These are just

$$\left(\frac{\partial u}{\partial x} - \frac{\partial v}{\partial y} \right)(x, y)$$

and

$$\left(\frac{\partial v}{\partial x} + \frac{\partial u}{\partial y} \right)(x, y)$$

so $\frac{\partial f}{\partial \bar{z}} = 0$ is indeed equivalent to the Cauchy-Riemann equations.

Next we show that conditions 3, 4, and 5 are equivalent. We can identify paths in Ω with paths in $\tilde{\Omega}$ by

$$\gamma(t) = x(t) + iy(t). \tag{8}$$

With this identification we have the relation

$$\int_C f(z)\,dz = \int (u\,dx - v\,dy) + i \int (v\,dx + u\,dy) \tag{9}$$

between the real and complex line integrals. It follows from Theorem U of the appendix that 3, 4, and 5 are equivalent.

Finally we prove that condition 1 implies 3 and condition 2 implies 1. Suppose condition 1 holds. The limit

$$\lim_{\Delta z \to 0} \frac{f(z + \Delta z) - f(z)}{\Delta z}$$

must be the same if we restrict Δz to real values Δx as if we restrict to purely imaginary values $i\Delta z$,

$$\lim_{\Delta x \to 0} \frac{f(z + \Delta x) - f(z)}{\Delta x} = \lim_{\Delta y \to 0} \frac{f(z + i\Delta y) - f(z)}{i\Delta y}. \tag{10}$$

Writing this equation in terms of u and v and taking the real and imaginary parts, we obtain the two equations

$$\lim_{\Delta x \to 0} \frac{u(x + \Delta x, y) - u(x, y)}{\Delta x} = \lim_{\Delta y \to 0} \frac{v(x, y + \Delta y) - v(x, y)}{\Delta y} \tag{11}$$

and

$$\lim_{\Delta x \to 0} \frac{v(x + \Delta x, y) - v(x, y)}{\Delta x} = - \lim_{\Delta y \to 0} \frac{u(x, y + \Delta y) - u(x, y)}{\Delta y} \tag{12}$$

This is exactly condition 3.

Now suppose condition 2 is satisfied. Let

$$g(t) = f(z + t\Delta z) \tag{13}$$

for real t. By the chain for partial derivatives

$$\begin{aligned}
\frac{dg}{dt}(t) &= \frac{d}{dt}(u + iv)(x + \frac{1}{2}\Delta z + \frac{1}{2}\overline{\Delta z}, y - \frac{i}{2}\Delta z + \frac{i}{2}\overline{\Delta z}) \tag{14}\\
&= \frac{\partial f}{\partial z}(z + t\Delta z)\Delta z + \frac{\partial f}{\partial \overline{z}}(z + t\Delta z)\overline{\Delta z}.
\end{aligned}$$

By condition 2 we can abbreviate this to

$$\frac{dg}{dt}(t) = \frac{\partial f}{\partial z}(z + t\Delta z)\Delta z. \tag{15}$$

By the fundamental theorem of calculus

$$\frac{f(z + \Delta z) - f(z)}{\Delta z} = \frac{g(1) - g(0)}{\Delta z} = \int_0^1 \frac{\partial f}{\partial z}(z + t\Delta z)\,dt \tag{16}$$

Theorem D of the appendix then shows that

$$\lim_{\Delta z \to 0} \frac{f(z + \Delta z) - f(z)}{\Delta z} = \frac{\partial f}{\partial z}(z). \tag{17}$$

This establishes condition 1 and completes the proof of the theorem.

The following theorem lists some easy consequences of criterion 1 for analyticity.

THEOREM 10 (USEFUL PROPERTIES OF ANALYTIC FUNCTIONS)
Let f and g be analytic functions on Ω.

1. *If $a, b \in \mathbb{C}$ then $af + bg$ is analytic on Ω.*

2. *fg is analytic on Ω.*

3. *If g has no zeroes in Ω then f/g is analytic on Ω.*

4. *If $U \subset \Omega$ is open then f restricted to U, written $f|_U$, is analytic on U.*

5. *If $\Omega = \cup_\alpha U_\alpha$ and $h|_{U_\alpha}$ is analytic on U_α for each α then h is analytic on Ω.*

Exercises

1. Show that if f is analytic then u and v are *harmonic*, in other words that they are twice continuously differentiable and satisfy Laplace's equation

$$\frac{\partial^2 u}{\partial x^2} + \frac{\partial^2 u}{\partial y^2} = 0, \qquad \frac{\partial^2 v}{\partial x^2} + \frac{\partial^2 v}{\partial y^2} = 0$$

   if f is an analytic function. Is the converse true?

2. Prove that $f(z) = e^z$ is analytic on $\mathbb{C}$. You can use

$$e^z = e^{\operatorname{Re} z}(\cos \operatorname{Im} z + i \sin \operatorname{Im} z)$$

   as the definition of e^z for complex z.

3. Prove that the function $f(z) = |z|$ is *not* analytic on any nonempty open set.

4. Prove that the composition of analytic functions is analytic. Specifically, show that if $f: \Omega_1 \to \Omega_2$ and $g: \Omega_2 \to \mathbb{C}$ are analytic then $g \circ f: \Omega_1 \to \mathbb{C}$ is analytic.

5. Prove the inverse function theorem: If $f: \Omega_1 \to \Omega_2$ is analytic and bijective and f' has no zeroes in Ω_1, then $f^{-1}: \Omega_2 \to \Omega_1$ is analytic.

6. The inverse function theorem remains valid even if we remove the hypothesis that f' has no zeroes. Try to prove it without this hypothesis. Don't be surprised if you are unsuccessful.

7. Prove that $g(z) = \log z$ is analytic on the complement of the negative real axis. You can take the properties

$$e^{\log z} = z$$

   and

$$-\pi < \operatorname{Im} \log z < \pi$$

as the definition of the logarithm for complex z. The second of these identifies what is commonly called the *principal branch* of the logarithm. No meaning is assigned to $\log z$ for z on the negative real axis. We define the *argument* of z by

$$\arg z = \operatorname{Im} \log z.$$

8. Prove that $h(z) = z^{\alpha}$ is analytic on the complement of the negative real axis. You can take

$$z^{\alpha} = e^{\alpha \log z}$$

as the definition of z^{α} for α and z complex. This is called the *principal branch* of the function z^{α}. No meaning is assigned to z^{α} for z on the negative real axis.

9. With z^{α} defined as above, does the identity

$$(xy)^{\alpha} = x^{\alpha} y^{\alpha}$$

hold for all x, y and α for which both sides are defined?

2.3   The Cauchy Integral Formula

The main result of this section is the following theorem, first stated and proved by Cauchy in [7].

THEOREM 11 (THE CAUCHY INTEGRAL FORMULA)
Let $\Omega \subset \mathbf{C}$ be open, f an analytic function on Ω, $z \in \Omega$, and C a closed contractible contour in Ω, of index 1 with respect to z. Then

$$f(z) = \frac{1}{2\pi i} \int_C \frac{f(\zeta)\, d\zeta}{\zeta - z}. \tag{1}$$

There is a useful extension of this theorem for derivatives. Suppose f, C, and z are as above. Formal differentiation under the integral sign in (1) gives

$$\frac{\partial^k f}{\partial z^k} = \frac{k!}{2\pi i} \int_C \frac{f(\zeta)\, d\zeta}{(\zeta - z)^{k+1}}. \tag{2}$$

The integral converges locally uniformly, so the differentiation under the integral sign is justified. We now proceed to the proof of the theorem.

Proof.   If we knew in advance that the function

$$g(\zeta) = \frac{f(\zeta) - f(z)}{\zeta - z} \tag{3}$$

were analytic in Ω then the proof would be easy. By criterion 2 for analyticity, we would have

$$\frac{1}{2\pi i} \int_C g(\zeta)\,d\zeta = 0 \tag{4}$$

or, equivalently,

$$f(z)\frac{1}{2\pi i} \int_C \frac{d\zeta}{\zeta - z} = \frac{1}{2\pi i} \int_C \frac{f(\zeta)\,d\zeta}{\zeta - z}. \tag{5}$$

By assumption the index is 1, so

$$f(z) = \frac{1}{2\pi i} \int_C \frac{f(\zeta)\,d\zeta}{\zeta - z}. \tag{6}$$

How can we prove that g is analytic on Ω? From parts 3 and 4 of Theorem 10 we know that g is analytic on $\Omega - \{z\}$. By criterion 1 for analyticity, applied to f, we know that g can be extended to a continuous function on Ω. The following theorem would then complete the proof.

THEOREM 12 (RIEMANN REMOVABLE SINGULARITIES THEOREM)
Let $\Omega \subset \mathbf{C}$ be open, $z \in \Omega$. Suppose h is continuous on Ω and analytic on $\Omega - \{z\}$. Then h is analytic on Ω.

This theorem is our first clue that the theory of analytic functions differs radically from that of differentiable functions. Unfortunately we will not be be able to prove Theorem 12 until we have at least a weak form of Theorem 11 available. The way out of this trap is as follows. First we will prove a weak version of the Cauchy integral formula, valid only when C is a small circle containing z. We will then complete the proofs of Theorems 11 and 12 on pages 91 and 93, respectively.

We use the notation

$$B(w, r) = \{z \in \mathbf{C} : |z - w| < r\} \tag{7}$$

and

$$\overline{B(w, r)} = \{z \in \mathbf{C} : |z - w| \le r\} \tag{8}$$

for the open and closed discs of radius r about w.

THEOREM 13 (WEAK CAUCHY)
Let Ω be an open subset of $\mathbf{C}$, f an analytic function on Ω, and W a point in Ω. Suppose $\overline{B(w, r)} \subset \Omega$. For any $z \in B(w, r)$ we have the representation

$$f(z) = \frac{1}{2\pi i} \int_{C'} \frac{f(\zeta)\,d\zeta}{\zeta - z}, \tag{9}$$

where C' is the contour corresponding to the path

$$\gamma'(t) = w + re^{2\pi i t}. \tag{10}$$

Proof. It suffices to check that

$$\left| \frac{1}{2\pi i} \int_{C'} \frac{f(\zeta)\,d\zeta}{\zeta - z} - f(z) \right| \leq \epsilon \tag{11}$$

for all $\epsilon > 0$. Let g be given by equation (3). By hypothesis f is analytic on Ω. Criterion 1 for analyticity shows that there are positive M and R such that

$$|g(\zeta)| = \left| \frac{f(\zeta) - f(z)}{\zeta - z} \right| \leq M \tag{12}$$

for $\zeta \in B(z, R)$. Let

$$0 < \rho < \min\left(R, r - |z - w|, \frac{\epsilon}{M}\right) \tag{13}$$

and let C'' be the contour corresponding to the path

$$\gamma''(t) = z + \rho e^{2\pi i t}. \tag{14}$$

C' and C'' are homotopic in $\omega - \{z\}$. Indeed

$$\sigma(t) = s\gamma''(t) + (1 - s)\gamma'(t) \tag{15}$$

is a homotopy. g is analytic on $\Omega - \{z\}$. Criterion 5 for analyticity shows that

$$\frac{1}{2\pi i} \int_{C'} g(\zeta)\,d\zeta = \frac{1}{2\pi i} \int_{C''} g(\zeta)\,d\zeta. \tag{16}$$

Inequality (12) of Section 2.1 shows that

$$\left| \frac{1}{2\pi i} \int_{C''} g(\zeta)\,d\zeta \right| \leq M\rho < \epsilon. \tag{17}$$

C' is of index 1 with respect to w, so

$$\frac{1}{2\pi i} \int_{C'} g(\zeta)\,d\zeta = \frac{1}{2\pi i} \int_{C'} \frac{f(\zeta)\,d\zeta}{\zeta - z} - f(z). \tag{18}$$

Combining (16), (17), and (18) gives (11), so the proof of theorem13 is complete.

We can now finish the proof of Theorem 11.

Proof. What remains to be shown is that g is analytic on Ω. We already know that G is analytic on $\Omega - \{z\}$. By part 5 of Theorem 10 it suffices to prove that g is analytic in $B(z, r)$ for some $r > 0$. Let r be sufficiently small that $\overline{B(z, r)} \subset \Omega$. Let $z' \in B(z, r)$. By Theorem 13,

$$g(z') = \frac{f(z') - f(z)}{z - z'} = \frac{1}{2\pi i} \int_{C} \frac{f(\zeta)\,d\zeta}{(\zeta - z')(\zeta - z)}, \tag{19}$$

where C is the circle of radius r about z, oriented counterclockwise. Formal differentiation under the integral sign is justified by Theorem S of the appendix. Thus g is analytic on $B(z, r)$ by criterion 2, and the proof of Theorem 11 is complete.

The argument given above for the analyticity of g in $B(z, r)$ can be generalized considerably, as the following theorem shows.

THEOREM 14 (ANALYTICITY OF INTEGRALS)

Let (a, b) be an interval, possibly infinite or semi-infinite, and Ω an open subset of $\mathbf{C}$. Let g be a function on $(a, b) \times \Omega$ such that

1. For any fixed $t \in (a, b)$ $g(t, z)$ is an analytic function on Ω.

2. The integral

$$f(z) = \int_a^b g(t, z)\, dt \tag{20}$$

converges locally uniformly in z.

Then f is analytic on Ω.

Proof. We will use the convergence criteria (2) and (3) of the appendix. The local uniform convergence of the integral (20) means that there is, for each $z_0 \in \Omega$, an $r > 0$ such that the integral converges uniformly for $|z - z_0| \leq r$. Thus there are, for any $\epsilon > 0$, $A(\epsilon)$ and $B(\epsilon)$ in the interval (a, b) such that

$$\left| \int_{\alpha'}^{\alpha''} g(t, \zeta)\, dt \right| \leq \epsilon \tag{21}$$

$$\left| \int_{\beta'}^{\beta''} g(t, \zeta)\, dt \right| \leq \epsilon \tag{22}$$

whenever $\zeta - z_0 \leq r$, $a < \alpha' \leq \alpha'' \leq A(\epsilon)$ and $B(\epsilon) \leq \beta' \leq \beta'' < b$. Formal differentiation under the integral sign in (20) gives

$$\frac{\partial f}{\partial z} = \int_a^b \frac{\partial g}{\partial z}(t, z)\, dt \tag{23}$$

and

$$\frac{\partial f}{\partial \bar{z}} = \int_a^b \frac{\partial g}{\partial \bar{z}}(t, z)\, dt = \int_a^b 0\, dt = 0. \tag{24}$$

We can justify this using Theorem S of the appendix provided we can show that both integrals converge uniformly in a disc about z_0. We will show this for the disc of radius $|z - z_0| \leq \frac{r}{2}$. The uniform convergence of (24) is obvious. Using equation (2) we can write

$$\int_{\alpha'}^{\alpha''} \frac{\partial g}{\partial z}(t, z)\, dt = \frac{1}{2\pi i} \int_{\alpha'}^{\alpha''} \int_C (\zeta - z)^{-2} g(t, \zeta)\, d\zeta\, dt. \tag{25}$$

By Theorem I of the appendix we have

$$\int_{\alpha'}^{\alpha''} \frac{\partial g}{\partial z}(t, z)\, dt = \frac{1}{2\pi i} \int_C (\zeta - z)^{-2} \int_{\alpha'}^{\alpha''} g(t, \zeta)\, dt\, d\zeta. \tag{26}$$

If $a < \alpha' \leq \alpha'' \leq A(\frac{2\epsilon}{r})$ then, by (21),

$$\left| (\zeta - z)^{-2} \int_{\alpha'}^{\alpha''} g(t, \zeta)\, dt \right| \leq \frac{r\epsilon}{2} \tag{27}$$

and the estimate (12) of Section 2.1 shows that

$$\left| \int_{\alpha'}^{\alpha''} \frac{\partial g}{\partial z}(t, z)\, dt \right| \leq \epsilon. \tag{28}$$

A similar argument shows that if $|z - z_0| \leq \frac{r}{2}$ and $B(\frac{2\epsilon}{r}) \leq \beta' \leq \beta'' < b$ then

$$\left| \int_{\alpha'}^{\beta''} \frac{\partial g}{\partial z}(t, z)\, dt \right| \leq \epsilon. \tag{29}$$

This establishes the uniform convergence of the integral (23) on the disc $|z - z_0| \leq \frac{r}{2}$ and completes the proof of Theorem 14.

As an application of Theorem 14 we now prove Theorem 12.

Proof. It suffices to show that h is analytic in a disc about z. Let r be sufficiently small that $\overline{B(z, r)} \subset \Omega$. We will show that

$$h(a) = \frac{1}{2\pi i} \int_{C'} \frac{h(\zeta)\, d\zeta}{\zeta - a} \tag{30}$$

for all $a \in B(z, r) - \{z\}$, where C' is contour corresponding to the path

$$\gamma'(t) = z + re^{2\pi i t}. \tag{31}$$

Note that this does not follow immediately from 11 because C' is not contractible in $\Omega - \{z\}$. Instead we proceed as follows. Let C'' be the path corresponding to the contour

$$\gamma''(t) = a + \rho e^{2\pi i t} \tag{32}$$

where $\rho > 0$ is chosen sufficiently small that $z \notin \overline{B(a, \rho)}$. Theorem 13 shows that

$$h(a) = \frac{1}{2\pi i} \int_{C''} \frac{h(\zeta)\, d\zeta}{\zeta - a}. \tag{33}$$

The integrand is analytic in $\Omega - \{z, a\}$. C' and C'' are not homotopic in $\Omega - \{z, a\}$, but they are homotopic in $\Omega - \{a\}$. Indeed

$$\sigma(s, t) = s\gamma''(t) + (1 - s)\gamma'(t) \tag{34}$$

is a homotopy. Let C_s be the contour corresponding to the path

$$\gamma_s(t) = \sigma(s, t). \tag{35}$$

There is an s_0 such that $z \in C_{s_0}$. In fact

$$s_0 = \frac{|z - a| - \rho}{|z - a| + r - \rho}, \tag{36}$$

but the precise value is irrelevant. What matters is that C_s is homotopic to C' in $\Omega - \{z, a\}$ for $s < s_0$ and homotopic to C' for $s > s_0$. Condition 5 of Theorem 9 therefore shows that

$$\int_{C_s} \frac{h(\zeta)\, d\zeta}{\zeta - a} = \int_0^1 \frac{h(\sigma(s, t))\, dt}{\sigma(s, t) - a} \tag{37}$$

is equal to $\int_{C'} \frac{h(\zeta)\, d\zeta}{\zeta - a}$ for $s < s_0$ and $\int_{C''} \frac{h(\zeta)\, d\zeta}{\zeta - a}$ for $s > s_0$. On the other hand Theorem G of the appendix shows that the right hand side is a continuous function of s, so

$$\int_{C'} \frac{h(\zeta)\, d\zeta}{\zeta - a} = \int_{C'} \frac{h(\zeta)\, d\zeta}{\zeta - a} \tag{38}$$

and equation (30) follows from (33).

Once we have equation (30) the proof is quite easy. By Theorem 14 the right hand side of (30) is an analyic function of a for all $a \in B(z, r)$, including $a = z$. This function agrees with $h(a)$ for all $a \neq z$ and h is continuous, so it agrees with $h(a)$ also for $a = z$, and the proof of Theorem 12 is complete.

The main idea of the proof of Theorem 14, using the formula (2) to convert a question about derivatives into a question about integrals, is also used to prove the following theorem.

THEOREM 15 (ANALYTICITY OF LIMITS)
Let $\Omega \subset \mathbf{C}$ be open and f_n a locally uniformly convergent sequence of functions analytic on Ω. Then their limit f is an analytic function and the successive derivatives of the f_n converge locally uniformly to those of f.

The statement about derivatives represents the first useful work we have managed to extract from the notion of analyticity. We have already proved by hand that various functions are smooth. For example, in Section 1.8 we proved that $\log \zeta(s)$ is smooth for $s > 1$. We did this by finding a representation for each successive derivative and verifying uniform convergence on an arbitrary closed subinterval. Now we have a much easier method available. In that example it would require us to check only the uniform convergence of the Euler product formula on certain discs. We would need to check this in a complex region, but this is, in fact, no more difficult than checking convergence for real values of s. The derivatives then come for free.

Proof. The statement about derivatives is an immediate consequence of equation (2). Thus f is continuously differentiable. We use criterion 2 for analyticity.

$$\int_C (\lim_{n \to \infty} f_n(z)) \, dz = \lim_{n \to \infty} \int_C f_n(z) \, dz = 0, \tag{39}$$

the interchange of the limit and integral can be justified as follows. Using the definition of the contour integral

$$\int_C (\lim_{n \to \infty} f_n(z)) \, dz = \int_a^b (\lim_{n \to \infty} f_n(\gamma(t))) \frac{d\gamma}{dt}(t) \, dt \tag{40}$$

$$= \int_a^b \lim_{n \to \infty} f_n(\gamma(t)) \frac{d\gamma}{dt}(t) \, dt \tag{41}$$

while

$$\lim_{n \to \infty} \int_C f_n(z) \, dz = \lim_{n \to \infty} \int_a^b f_n(\gamma(t)) \frac{d\gamma}{dt}(t) \, dt. \tag{42}$$

These are equal by Theorem D of the appendix.

Exercises

1. Prove that

$$f(t) = \int_0^\infty t^{z-1} e^{-t} \, dt$$

   is analytic for $\text{Re } z > 0$.

2. Prove that

$$f(t) = \lim_{n \to \infty} \frac{n! n^z}{(z)_{n+1}}$$

   is analytic for $z \notin \{0, -1, -2, \cdots\}$.

3. Prove that

$$f(t) = \frac{e^{-\gamma z}}{z} \prod_{n=1}^{\infty} (1 + \frac{z}{n})^{-1} e^{z/n}$$

   is analytic for $z \notin \{0, -1, -2, \cdots\}$.

4. Prove that $f(s) = (s - 1)\zeta(s)$ is analytic everywhere.

5. Suppose f is analytic in the disc of radius ρ about z. Use the Cauchy integral form to prove the estimate

$$|f(z_0)| \leq \max_{|z-z_0|=r} |f(z)|,$$

   valid for $0 \leq r < \rho$.

6. Suppose f is analytic in the disc of radius ρ about z. Prove the *Cauchy estimates*

$$|f^{(n)}(z_0)| \leq n! r^{-n} \max_{|z-z_0|=r} |f(z)|,$$

   valid for $0 \leq r < \rho$ and n a nonnegative integer.

7. *Liouville's theorem* states that a bounded function analytic on all of $\mathbf{C}$ must be constant. Prove this theorem. Hint: see the preceding exercise.

8. Apply the Cauchy estimates to $f(z) = e^z$, $z_0 = 0$. The point is not to get upper bounds on 1 but to get lower bounds on $n!$. The optimal choice of r depends on n.

9. The Riemann removable singularities theorem has the following analogue in $\mathbf{C}^2$. If $h: \mathbf{C}^2 - \{0\}$ is continuous and, for each fixed value of z_1, analytic in z_2, and vice versa, then h can be extended to a function on $\mathbf{C}^2$ with the same properties. Note that we do not need to assume that h is continuous at 0. Prove this theorem.

10. Show, by means of an example, that the hypothesis that h is continuous at the singularity cannot be removed from the one-dimensional version of the Riemann removable singularities theorem.

2.4   Power Series and Rigidity

We will see in this section that the analytic functions are precisely those which can be represented locally by power series. This fact places strong restrictions on the behavior of analytic functions. The most important of these is *rigidity*, the fact that an analytic function which vanishes on a nonempty open subset of a connected set must vanish identically.

THEOREM 16 (POWER SERIES REPRESENTATION)
Let $\Omega \subset \mathbf{C}$ be open and f a function on Ω. The following two conditions are equivalent.

1. *f is analytic on Ω.*

2. *For every $w \in \Omega$ there is a neighborhood U_w of w and a sequence a_n, depending on w, such that*

$$f(z) = \sum_{n=0}^{\infty} a_n(z - w)^n \qquad (1)$$

for all $z \in U_w$.

Proof. Suppose condition 2 is satisfied. Let $x \in U_w - \{w\}$. Then

$$\sum_{n=0}^{\infty} a_n(x - w)^n \qquad (2)$$

converges so

$$|a_n||x - w|^n \le C \qquad (3)$$

with some constant C, depending on w, but not on n. If $0 < r < |x - w|$ then, for all $z \in \overline{B}(w, r)$, the sum

$$\sum_{n=0}^{\infty} a_n(z - w)^n \qquad (4)$$

converges uniformly, by comparison with the geometric series

$$C \sum_{n=0}^{\infty} \left(\frac{r}{|x - w|} \right)^n . \qquad (5)$$

Thus the sum (4) converges locally uniformly on $B(w, |x - w|)$. The partial sums are polynomials, and hence analytic. Theorem 15 shows that the series represents an analytic function on $B(w, |x - w|)$. Taking the union over all $x \in U_w$ and, using part 5 of Theorem 10, we see that the series represents an analytic function, at least on the smallest open disc about w containing U_w. Note, however, that the series may not be equal to f throughout that disc! We have shown, however, that f is analytic at least in U_w. Another application of part 5 of Theorem 10 shows that f is analytic on Ω, in other words that condition 1 is satisfied.

Next we suppose that condition 1 is satisfied and prove condition 2. Define

$$\rho = \sup\{r \in \mathbf{R} : \overline{B}(w, r) \subset \Omega\}. \qquad (6)$$

Since Ω is open, $\rho > 0$. Let $0 < r < \rho$ and let C be the circle of radius r about w, oriented counterclockwise. Expanding $(\zeta - z)^{-1}$ in powers of $z - w$

$$\frac{1}{\zeta - z} = \sum_{n=0}^{\infty} \frac{(z - w)^n}{(\zeta - w)^{n+1}} \tag{7}$$

in the Cauchy integral formula

$$f(z) = \frac{1}{2\pi i} \int_C \frac{f(\zeta)\, d\zeta}{\zeta - z} \tag{8}$$

gives the formula

$$f(z) = \sum_{n=0}^{\infty} \frac{(z - w)^n}{2\pi i} \int_C \frac{f(\zeta)\, d\zeta}{(\zeta - w)^{n+1}}. \tag{9}$$

The term by term integration is justified by uniform convergence for $|z - w| < r$. We can use equation (2) from the previous section to rewrite (9) as

$$f(z) = \sum_{n=0}^{\infty} \frac{f^{(n)}(w)}{n!} (z - w)^n. \tag{10}$$

This representation is valid for all $|z - w| < r$ and all $r < \rho$, and hence for all $|z - w| < \rho$. Thus condition 2 is satisfied with U_w the large open disc about w contained in Ω, and

$$a_n = f^{(n)}(w)/n!. \tag{11}$$

This formula is hardly surprising; nothing else would be compatible with Taylor's theorem. This completes the proof of Theorem 16.

The following easy corollary of Theorem 16 will be of great use in this section and the next.

THEOREM 17 (LOCAL FACTORIZATION)
Let $\Omega \subset \mathbf{C}$ be open, and f analytic on Ω. Every point w in Ω has a neighborhood U_w where either

1.

$$f(z) = 0 \tag{12}$$

identically, or

2.

$$f(z) = (z - w)^{v_w(f)} u(z) \qquad (13)$$

where $v_w(f)$ is the order of vanishing of f at w, and u is an analytic function with no zeroes in U_w. In particular, f has no zeroes in the punctured neighborhood $U_w - \{w\}$.

Proof. If $f^{(n)}(w) = 0$ for all n then the proof of Theorem 16 shows that (12) holds, at least throughout the largest disc about w contained in Ω. Otherwise there is an integer k such that $f^{(k)} \neq 0$, while $f^{(n)} = 0$ for $n < k$. This k is, of course, the order of vanishing of f at w, in other words the number we have called $v_w(f)$ above, but this looks ugly as an exponent, so we will continue to call it k. By Taylor's theorem

$$f(z) = f^{(k)}(w)(z - w)^k + o(|z - w|^k) \qquad (14)$$

for z near w. Defining u by

$$u(z) = (z - w)^{-k} f(z) \qquad (15)$$

for $z \in \Omega - \{w\}$, we find

$$u(z) = f^{(k)}(w) + o(1) \qquad (16)$$

for z near w. Theorem 10, part 3 shows that u is analytic on $\Omega - \{w\}$ and continuous at w. Theorem 12 shows that it is analytic on Ω. Equation (16) shows that $u(w) \neq 0$ and, by continuity, $u(z) \neq 0$ in some neighborhood U_w of w, so the proof of Theorem 17 is complete.

Our first application of Theorem 17 is the following.

THEOREM 18 (RIGIDITY PRINCIPLE)
Let Ω be an open, connected subset of $\mathbf{C}$ and f an analytic function on Ω. Either

1. $f(z) = 0$ for all $z \in \Omega$, or

2. the zeroes of f are all of finite order and discrete, i.e. each has a neighborhood containing no other zeroes.

Proof. Define

$$V_n = \{z \in \Omega : f^{(n)}(z) = 0\} \qquad (17)$$

and

$$V = \cap_{n=0}^{\infty} V_n. \qquad (18)$$

By Theorem 15 $f^{(n)}$ is continuous for all n, and thus V_n is closed. As the intersection of a collection of closed sets, V is closed as well. On the other hand, if $w \in V$ then Theorem 17 gives a whole neighborhood U_w of w which is contained in V, so V is open as well. Ω was assumed connected, so either $V = \Omega$ or $V = \emptyset$. In the former case $\Omega \subset V_0$, so 1 holds. In the latter case every zero is of finite order. Theorem 17 then shows that each of these zeroes has a neighborhood containing no others, and the proof of Theorem 18 is complete.

The rigidity principle is most often applied to the difference $f = f_1 - f_2$ of two analytic functions. It shows that if $f_1(z) = f_2(z)$ for all z in some set S with a point of accumulation in the connected open set Ω, then $f_1(z) = f_2(z)$ for all $z \in \Omega$. The name "rigidity" is derived from this phenomenon; analytic functions are rigid in the sense that they are uniquely determined by their restrictions to such a subset S.

As an example of the rigidity principle we consider the identity

$$\lim_{n \to \infty} \frac{n! n^z}{(z)_{n+1}} = z^{-1} e^{-\gamma z} \prod_{j=1}^{\infty} (1 + z/j)^{-1} e^{z/j}. \tag{19}$$

We proved this in Section 1.1 for $z \in S$, where

$$S = \{z \in \mathbf{R} : z > 0\}. \tag{20}$$

Both

$$\frac{n! n^z}{(z)_{n+1}}$$

and

$$z^{-1} e^{-\gamma z} \prod_{j=1}^{n} (1 + z/j)^{-1} e^{z/j}$$

are analytic on the connected open set

$$\Omega = \{z \in \mathbf{C} : z \notin \mathbf{Z} \text{ or } z > 0\}. \tag{21}$$

Both sequences converge locally uniformly on Ω, as the reader may easily check, so both the right and left hand sides in (19) are analytic functions on Ω by Theorem 15. Since S has limit points in Ω, (19) must hold for all $z \in \Omega$.

Though powerful, the rigidity theorem has two important limitations. The first is that it cannot be used to extend inequalities. Thus $\Gamma(z) > 0$ for all $z \in S$, but this is not true throughout Ω. One unfortunate consequence is that the rigidity principle cannot be used to extend asymptotic expansions. We cannot, for example, use the rigidity principle to remove the restriction $p > 0$ from the asymptotic expansion for the Whittaker integral $J(p, q, s)$ in Section 1.5. The second limitation, and trap for the unwary, is that the analyticity of f_1 and f_2 throughout Ω belongs to the hypotheses, not the conclusion, of Theorem 18. Thus the identity

$$\lim_{n \to \infty} \frac{n! n^z}{(z)_{n+1}} = \int_0^\infty t^{z-1} e^{-t} \, dt \tag{22}$$

is valid only for $\operatorname{Re} z > 0$, even though the left hand side is analytic throughout Ω.

Exercises

1. If

$$f(z) = \sum_{n=0}^{\infty} a_n (z - w)^n$$

   for $|z - w| < r$, what is the power series expansion of $\frac{\partial f}{\partial z}$? Is it convergent for $|z - w| < r$?

2. If

$$f(z) = \sum_{n=0}^{\infty} a_n (z - w)^n$$

   for $|z - w| < r$ and

$$f(z) = \sum_{n=0}^{\infty} b_n (z - w)^n$$

   for $|z - w| < r'$ what are the power series expansions $f(z) + g(z)$ and $f(z)g(z)$? For which z can you guarantee that they converge?

3. Suppose f and g have expansions as in the preceding exercise and that $g(w) = w$. How can you calculate the coefficients in the expansion of $f(g(z))$? Can you say anything intelligent about the radius of convergence?

4. Return to Exercise 2 of Section 1.4 and give an alternate proof.

5. Return to Exercise 7 of Section 1.5 and give a civilized proof.

6. Give an example of an open subset U of $\mathbf{C}$, a function f analytic on U, a point $w \in \omega$, and an open disc B, centered at w, such that

   (a) The Taylor series of f about w,

$$\sum_{n=0}^{\infty} a_n (z - w)^n$$

   converges throughout D.

(b) There are points $z \in D \cap U$ where

$$f(z) \neq \sum_{n=0}^{\infty} a_n(z-w)^n.$$

2.5   The Distribution of Primes III

We return to the problem of estimating

$$d(n) = \pi(n) - \operatorname{li} n. \tag{1}$$

We would like to prove the following strengthened form of Theorem 2.

> For any positive real ϵ and δ, there is a positive integer N such that the inequality
>
> $$|d(n)| \leq \delta n^{1/2+\epsilon}$$
>
> holds for all $n \geq N$.

Unfortunately we are unable to do so. Instead we prove a strengthened form of Theorem 6.

THEOREM 19
For any positive real ϵ and δ and any positive integer N the inequality

$$|d(n)| \leq \delta n^{1/2+\epsilon} \tag{2}$$

holds for some $n \geq N$.

   This can be improved. The ultimate result of this type is due to Littlewood, Ingham, and Skewes, who showed in a series of papers from 1914 to 1955 that $d(n)$ changes sign infinitely often. Skewes' proof, unlike the one we give below, even gives numerical bounds, though these are too large to be of practical value.[1] The proof of this statement is technically quite complicated. The reader may find it, together with more historical information, in Skewes' paper [62]. Our purpose here is not to obtain precise estimates on primes but to illustrate the power of the theorems developed in sections 2.2, 2.3, and 2.4. We will therefore be content with Theorem 19. At the end of the section we will make a few brief remarks concerning the proofs of the theorems of Littlewood, Ingham, and Skewes.

---

[1]For small n we have $d(n) < 0$. "Small" here includes all values for which $\pi(n)$ is known. From the work of Skewes it follows that $\pi(X) > \operatorname{li} X$ for some $X < 10^{10^{10^{34}}}$.

In Section 1.8 we showed that $\zeta(s)$ extends to a function which is smooth for all real $s \neq 1$ and that

$$u(s) = \sum_{n=2}^{\infty} n^{-s} - \sum_{p} p^{-s} \log p \tag{3}$$

extends to a function which is smooth for all real s such that $s > \frac{1}{2}$ and $\zeta(s) \neq 0$. The proofs involved showing that certain sums and integrals converge locally uniformly in s and then applying theorems O and S of the appendix. If, in place of those two theorems, we use theorems 15 and 14, we see that $\zeta(s)$ extends to a function which is analytic for all complex $s \neq 1$ and that $u(s)$ extends to a function which is analytic for all complex s such that $\text{Re}\, s > \frac{1}{2}$ and $\zeta(s) \neq 0$. The problem is to locate the zeroes of ζ. We will have more to say about this in Section 2.6. For now we will simply accept that $\zeta(s)$ is known to be nonzero in the region

$$\Omega = \{s \in \mathbf{C}\colon \text{Re}\, s > \frac{1}{2} \text{ and } |\text{Im}\, s| \leq 1\}, \tag{4}$$

so that u is analytic in Ω as is

$$v(s) = \int_{1}^{s} u(\sigma)\, d\sigma. \tag{5}$$

With this at our disposal we can prove Theorem 19.

Proof. Suppose Theorem 19 is false. Then there is an N_1 such that

$$|d(n)| \geq \delta n^{1/2+\epsilon} \tag{6}$$

for all $n \geq N_1$. Let N_2 be chosen such that

$$N_2 \geq \frac{1}{4}\delta^{-2} \tag{7}$$

and write

$$N = \max(N_1, N_2). \tag{8}$$

Then, for $n \geq N$, either

$$d(n) \geq \delta n^{1/2+\epsilon} \tag{9}$$

or

$$d(n) \leq -\delta n^{1/2+\epsilon}. \tag{10}$$

Similarly either

$$d(n+1) \geq \delta(n+\frac{1}{2})^{1/2+\epsilon} \tag{11}$$

or

$$d(n+1) \leq -\delta(n+\frac{1}{2})^{1/2+\epsilon}. \tag{12}$$

On the other hand we saw in Section 1.9 that

$$|d(n+1) - d(n)| \leq 1 \tag{13}$$

and hence, since $n \geq N_2$ and $\epsilon \geq 0$,

$$|d(n+1) - d(n)| \leq \delta n^{1/2+\epsilon} + (n+\frac{1}{2})^{1/2+\epsilon}. \tag{14}$$

It follows that (9) and (12) are incompatible as are (10) and (11). In other words either (9) holds for all $n > N$ or (10) holds for all $n > N$. We will show that (9) leads to a contradiction. The argument for (10) is the same except for some signs.

For any smooth function f and any $z \neq 1$ we have

$$\frac{1}{m!} \frac{d^m}{dz^m} \frac{f(w+sz)}{1-z} = \sum_{j=0}^{m} \frac{s^j}{j!} f^{(j)}(w+sz), \tag{15}$$

a formula which is easily proved by induction. For $z = 0$ we find

$$\frac{1}{m!} \frac{d^m}{dz^m}|_{z=0} \frac{f(w+sz)}{1-z} = \sum_{j=0}^{m} \frac{s^j}{j!} f^{(j)}(w). \tag{16}$$

If f is analytic the Cauchy estimates show that

$$|\sum_{j=0}^{m} \frac{s^j}{j!} f^{(j)}(w)| \leq r^{-m} \max_{|z| \leq r} |\frac{f(w+sz)}{1-z}|. \tag{17}$$

We will use (17) to estimate the sums on the right hand side in equation (38) of Section 1.9.

Let

$$s = 1 + \frac{\epsilon}{2} \tag{18}$$

and

$$r = \frac{1}{2s}. \tag{19}$$

We apply (17) to $f = \zeta$, $w = s + 1$. This produces

$$\sum_{j=0}^{m} \frac{s^j}{j!} \frac{d^j \zeta}{ds^j}(s+1) \leq r^{-m} \max_{|z| \leq r} \left| \frac{\zeta(1 + s + sz)}{1 - z} \right| \tag{20}$$

$$\leq 2r^{-m} \max_{\operatorname{Re} s \geq 3/2} |\zeta(1 + s + sz)|$$

$$= 2\zeta(\tfrac{3}{2})r^{-m}.$$

Equation (38) of Section 1.9 shows that

$$\left| \sum_{j=1}^{m} \frac{E(j, s)}{j!} s^j \right| \leq 2C\zeta(\tfrac{3}{2})(m + 1)(m + 2)r^{-m}. \tag{21}$$

Next we apply (17) to $f = v$, $w = s$. This gives

$$\left| \sum_{j=1}^{m} \frac{Y(j, s)}{j!} s^j \right| = \left| \sum_{j=1}^{m} \frac{s^j}{j!} \frac{d^j v}{ds^j}(s) \right| \tag{22}$$

$$\leq r^{-m} \max_{|z| \leq r} \left| \frac{v(s + sz) - v(s)}{1 - z} \right|.$$

Setting

$$K = 2C\zeta(\tfrac{3}{2}) + \max_{|z| \leq r} \left| \frac{v(s + sz) - v(s)}{1 - z} \right| + |sX(0, s)| \tag{23}$$

we see from equation (38) of Section 1.9 that

$$\left| \frac{X(m, s)s^m}{m!} \right| \leq K(m + 1)(m + 2)s^{-1}r^{-m}. \tag{24}$$

Let $t = \tfrac{1}{2} - \tfrac{\epsilon}{2}$. By the comparison test

$$\sum_{m=0}^{\infty} \left| \frac{X(m, s)t^m}{m!} \right| \leq Ks^{-1} \sum_{m=0}^{\infty} (m + 1)(m + 2)(\tfrac{t}{rs})^m \tag{25}$$

$$\leq Ks^{-1} \sum_{m=0}^{\infty} (m + 1)(m + 2)(1 - \epsilon)^m$$

$$\leq 4Ks^{-1}\epsilon^{-3}$$

is convergent, as is

$$\sum_{m=0}^{\infty} \frac{X(m, s)t^m}{m!} = \sum_{m=0}^{\infty} \sum_{n=2}^{\infty} \frac{d(n)n^{-s-1}t^m \log^m n}{m!}. \tag{26}$$

It is easy to check that

$$\sum_{m=0}^{\infty}\sum_{n=2}^{N}\frac{d(n)n^{-s-1}t^{m}\log^{m}n}{m!}$$

is convergent. Indeed

$$\sum_{m=0}^{\infty}\sum_{n=2}^{N}\frac{d(n)n^{-s-1}t^{m}\log^{m}n}{m!}=\sum_{n=2}^{N}\sum_{m=0}^{\infty}\frac{d(n)n^{-s-1}t^{m}\log^{m}n}{m!} \qquad (27)$$

$$=\sum_{n=2}^{N}d(n)n^{t-s-1}.$$

It follows that

$$\sum_{m=0}^{\infty}\sum_{n=N+1}^{\infty}\frac{d(n)n^{-s-1}t^{m}\log^{m}n}{m!}$$

converges. All terms in the series are positive, so we may reverse the order of summation.

$$\sum_{n=N+1}^{\infty}\sum_{m=0}^{\infty}\frac{d(n)n^{-s-1}t^{m}\log^{m}n}{m!}$$

therefore converges. But this is just

$$\sum_{n=N+1}^{\infty}d(n)n^{t-s-1}$$

or

$$\sum_{n=N+1}^{\infty}d(n)n^{3/2-\epsilon}.$$

By equation (9) this is bounded from below by the divergent series

$$\delta\sum_{n=N+1}^{\infty}d(n)n^{-1}.$$

This contradiction shows that the hypothesis (9) is untenable. The proof of the theorem is therefore complete.

We should note that the convergence of the series

$$\sum_{n=1}^{\infty}d(n)n^{3/2-\epsilon}$$

was proved only under the assumption that (9) holds for all sufficiently large n, an assumption subsequently shown to be invalid. It is not known at present whether this series is convergent for all positive ϵ.

Theorem 19 was proved using the Cauchy estimates. In other words it was proved by estimating a contour integral over a circle. That circle was contained in the region Ω. But the function u is analytic in a larger region. Pushing the contour of integration further gives better estimates. This, however, requires more information on the zeroes of zeta. Riemann conjectured that $\zeta(s) \neq 0$ when $\operatorname{Re} s > \frac{1}{2}$. Under this assumption Littlewood proved that $d(n)$ changes sign infinitely often. Riemann's conjecture remains unproved, but Skewes showed how to prove, even without this hypothesis, that $d(n)$ changes sign infinitely often. This involves moving the contour past potential zeroes in the region $\operatorname{Re} s > \frac{1}{2}$. The required estimates are rather complicated, and we will not pursue the matter further here.

2.6 Meromorphic Functions

If g and h are analytic on Ω then, as we have seen in Theorem 10, $f = g/h$ is analytic in the complement of the zeroes of h. With Theorem 17 available, we know enough about the zeroes of h to say rather more.

THEOREM 20 (EQUIVALENT DEFINITIONS OF MEROMORPHICITY)
Let Ω be an open subset of $\mathbf{C}$ and f a function defined on a subset of Ω. The following conditions are equivalent.

1. *Every point $w \in \Omega$ has a punctured neighborhood $U_w - \{w\}$ where f is defined and can be represented as a quotient*

$$f(z) = g_w(z)/h_w(z), \tag{1}$$

   where the functions g_w and h_w are analytic on U_w.

2. *Every point $w \in \Omega$ has a punctured neighborhood $U_w - \{w\}$ where f is defined and can be represented by an infinite series,*

$$f(z) = \sum_{n=v_w(f)}^{\infty} a_n(z - w)^n \tag{2}$$

   with some integer $v_w(f)$ and a sequence a_n, depending on w but not on z, with $a_{v_w(f)} \neq 0$. This series is called the Laurent expansion *of f about the point w.*

3. *Every point $w \in \Omega$ has a punctured neighborhood $U_w - \{w\}$ where f is defined and can be represented as*

$$f(z) = (z - w)^{v_w(f)} u(z) \tag{3}$$

with some integer $v_w(f)$ and a function u, analytic on U_w with no zeroes there.

All of these criteria are "local." In one respect this is odd. Analytic functions are, in some sense, a generalization of polynomials. Meromorphic functions are then a generalization of rational functions. One might therefore expect a definition of meromorphic functions like

THEOREM 21
A function f on an open subset U of $\mathbf{C}$ is meromorphic if and only if there exist analytic functions g and h on U such that the zeroes of h are discrete and

$$f(z) = g(z)/h(z), \tag{4}$$

except where $h(z) = 0$.

This would give a "global" criterion for analyticity. There are many reasons to avoid this approach though. Global criteria are often difficult to verify. It is not easy, for example, to show that the Gamma function satisfies the condition of Theorem 21. By contrast it is quite easy to show that it satisfies criterion 1 of Theorem 20. Indeed, if Re $w > -n$ then the functional equation

$$\Gamma(z) = \frac{\Gamma(z + n)}{(z)_n} \tag{5}$$

gives a representation of the form (1) on

$$U_w = \mathbf{C} - \{-n, , -n - 1, \ldots\}. \tag{6}$$

There is no such simple way to get a representation of the form (4).

Another reason to prefer a local definition is that it makes the proofs of many theorems easier. We have used part 5 of Theorem 10 many times. The corresponding property of meromorphic functions is obvious, if we take the criteria of Theorem 20 as the definition of meromorphicity. If we take the criterion of Theorem 21 as our definition then this theorem becomes much harder to prove.

Finally, we would like a definition which generalizes nicely. We will not discuss Riemann surfaces in this book, but a few quick remarks may clarify the distinction between our local and global criteria. A *Riemann surface* is just a one-dimensional complex manifold. In other words, it can be equipped with coordinate charts in such a way that the transition functions are analytic on the overlaps.

The most interesting Riemann surfaces are those which are compact. These are all algebraic curves over $\mathbf{C}$, though this requires a fair amount of effort to prove. If we were to define meromorphic functions on a Riemann surface by Theorem 21 we would get only the constant functions. If we define them by the criteria of Theorem 20 we get precisely the field of rational functions on our algebraic curve. Similar remarks apply to the generalization to domains in $\mathbf{C}^n$. If we were to adopt Theorem 21 as our definition we would get an uninteresting class of functions. In this case criterion 1 of Theorem 20 gives the "correct" class of functions.

For all of these reasons we choose to define a meromorphic function as one which satisfies any of the conditions of Theorem 20. With this choice Theorem 21 is a theorem, rather than a definition. We need not bother to prove it until we need it. In fact we will never need it, so we will never prove it.

We stated above that meromorphic functions are, in some sense, a generalization of rational functions. The most useful sense in which this is true is that the residue theorem, Theorem 7 of Section 2.1, can be extended to meromorphic functions. That extension is the main goal of this section. It requires, of course that we extend the definitions of pole, order, and residue to meromorphic functions. This is quite easy. The points where $v_w(f) < 0$ are called *poles*, the integer $-v_w(f)$ the *order* of the pole, and the coefficient a_{-1} the *residue* of f at the pole w, denoted $\text{Res}_{z=w} f(z)$. We note that rational functions are meromorphic by criterion 1 and that the definitions of pole, order, and residue given above agree with those of Section 2.1. It is traditional to extend f to a function from all of Ω to $C \cup \{\infty\}$ by assigning it the value ∞ at the poles and using the power series (2) at any other points w where f is not initially defined. We must still prove Theorem 20. Fortunately most of the work was done in the preceding section.

Proof. Suppose condition 1 is satisfied. Theorem 17 gives the local factorizations

$$g_w(z) = (z - w)^{v_w(g)} u_{g,w}(z) \tag{7}$$

and

$$h_w(z) = (z - w)^{v_w(h)} u_{h,w}(z) \tag{8}$$

with $u_{g,w}$ and $u_{h,w}$ analytic and nonzero in a neighborhood U_w of w. This gives a representation of the form (3) with

$$v_w(f) = v_w(g) - v_w(h) \tag{9}$$

and

$$u(z) = u_{g,w}(z)/u_{h,w}(z). \tag{10}$$

u is analytic by part 3 of Theorem 10. This establishes 3.

Suppose next that condition 3 is satisfied. Theorem 16 allows us to expand u in power series,

$$u(z) = \sum_{n=0}^{\infty} c_n (z - w)^n, \tag{11}$$

where $c_0 = u(w) \neq 0$, so we have a representation of the form (2) with

$$a_n = c_{n-v_w(f)} \tag{12}$$

and $a_{v_w(f)} = c_0 \neq 0$. Thus condition 2 is satisfied.

Suppose finally that 2 is satisfied. If $v_w(f) \leq 0$ then f extends to an analytic function on U_w by Theorem 16. In this case we may take $g = f$ and $h = 1$. Otherwise set

$$g(z) = \sum_{n=0}^{\infty} a_{n+v_w(f)} (z - w)^n. \tag{13}$$

The convergence of (2) implies that of (13), so Theorem 16 shows that g is analytic for z near w. We may therefore take g as above, and $h(z) = (z - w)^{-v_w(f)}$. Thus condition 1 is satisfied, and the proof of Theorem 20 is complete.

THEOREM 22 (RESIDUE THEOREM FOR MEROMORPHIC FUNCTIONS)
Let $\Omega \subset \mathbf{C}$ be open and f a meromorphic function on Ω. If C is a closed contractible contour in Ω then

$$\frac{1}{2\pi i} \int_C f(z)\, dz = \sum_{\zeta} \mathrm{Ind}(C, \zeta) \operatorname*{Res}_{z=\zeta} f(z) \tag{14}$$

where the sum is taken over the poles of f. In words, the contour integral is $2\pi i$ times the sum of the residues weighted by the indices.

Proof. The contractibility hypothesis gives a function σ from $[0, 1] \times [0, 1]$ to Ω with

$$\begin{aligned}
\sigma(1, t) &= \gamma(t), \\
\sigma(s, 1) &= \sigma(s, 0), \\
\sigma(0, t) &= \text{constant}.
\end{aligned} \tag{15}$$

Let K denote the image of $[0, 1] \times [0, 1]$ under the map σ. Our first task is to show that there are only finitely many poles of f in K. Suppose on the contrary that there are infinitely many. In this case we form a sequence w_j of distinct poles in K. Since $w_j \in K$ there must be $0 \leq s_j, t_j \leq 1$ such that

$$w_j = \sigma(s_j, t_j). \tag{16}$$

Theorem J of the appendix shows that there is a subsequence (s_{j_k}, t_{j_k}) such that

$$\lim_{k \to \infty} (s_{j_k}, t_{j_k}) = (S, T) \qquad (17)$$

for some $0 \leq S, T, \leq 1$. By the continuity of σ

$$\lim_{k \to \infty} w_{j_k} = \sigma(S, T) \in \Omega. \qquad (18)$$

But this is impossible since, by criterion 3 for meromorphicty, $\lim_{k \to \infty} w_{j_k}$ has a punctured neighborhood which contains no poles of f. Thus the assumption that f has infinitely many poles in K is untenable.

Next we show that we can write f in the form

$$f(z) = r(z) + g(z), \qquad (19)$$

where r is rational and g is analytic in some open set U containing K. Let $w \in K$ be arbitrary. Criterion 2 for meromorphicity together with Theorem 1 shows that

$$f(z) - \sum_{\substack{v_w(f) + n < 0 \\ n > 0}} a_n(z - w)^n$$

is analytic in a neighborhood of w. Since there are only finitely many poles of f in K, the sum

$$\sum_{\substack{\zeta \in K - \{w\} \\ v_\zeta(f) + n < 0 \\ n > 0}} a_n(z - \zeta)^n,$$

is also analytic in a neighborhood of w. Call the intersection of these two neighborhoods U_w. In U_w the function $g = f - r$, where r is defined by

$$r(z) = \sum_{\substack{\zeta \in K \\ v_\zeta(f) + n < 0 \\ n > 0}} a_n(z - \zeta)^n, \qquad (20)$$

is analytic. By Theorem 10, part 5, g is analytic throughout

$$U = \cup_{w \in K} U_w. \qquad (21)$$

Using again the fact that f has only finitely many zeroes in K, we see that r is a rational function.

The contour C is also contractible in U. Indeed the same homotopy σ will work. g is analytic in U, so criterion 2 for analyticity gives the equation

$$\frac{1}{2\pi i} \int_C g(z)\, dz = 0. \tag{22}$$

Theorem 7 gives the equation

$$\frac{1}{2\pi i} \int_C r(z)\, dz = \sum_\zeta \text{Ind}(C, \zeta) \operatorname*{Res}_{z=\zeta} r(z) \tag{23}$$

where the sum is over the poles of r in U. Adding these two equations, and recalling that the poles and residues of f and r in U are the same, we get the equation

$$\frac{1}{2\pi i} \int_C f(z)\, dz = \sum_\zeta \text{Ind}(C, \zeta) \operatorname*{Res}_{z=\zeta} f(z), \tag{24}$$

where the sum is taken over the poles of f in U. This agrees with equation (14), except that the latter had a sum over all poles of f in Ω.

If ζ is a pole outside of U then $(z - \zeta)^{-1}$ is an analytic function of z on U, so another application of criterion 2 for analyticity gives the equation

$$\frac{1}{2\pi i} \int_C \frac{dz}{z - \zeta} = 0. \tag{25}$$

The quantity on the left is the index of C with respect to ζ, so all the terms in the sum on the right of equation (14) corresponding to poles outside of U are zero, and Theorem 22 is proved.

One special case of Theorem 22 is important enough to merit special mention.

THEOREM 23 (ZERO COUNTING THEOREM)
Let $\Omega \subset \mathbf{C}$ be open and f an analytic function on Ω. If C is a closed contractible contour in Ω then

$$\frac{1}{2\pi i} \int_C \frac{\partial}{\partial z} \log f(z) = \sum_\zeta \text{Ind}(C, \zeta) v_\zeta(f) \tag{26}$$

where $v_\zeta(f)$ is the order of vanishing of f at ζ and the sum is taken over the zeroes of f.

In general $\log f$ is not well defined on on Ω, but

$$\frac{\partial}{\partial z} \log f(z) = \frac{\frac{\partial f}{\partial z}(z)}{f(z)} \tag{27}$$

is a well-defined meromorphic function on Ω. This contour integral can be used to count the zeroes, with multiplicities, in a region R. One simply takes C to be a contour surrounding R oriented counterclockwise and computes the integral. Since the answer is known to be an integer, one can do a numerical calculation with accuracy $1/2$ and round off. A slightly more sophisticated version of this idea is used to locate the zeroes of the zeta function.

We will use Theorem 23 twice. Our first application will be to the fundamental theorem of algebra later in this section. Our second application will be to locate the zeroes of theta functions in Section 3.1.

Proof. Let $w \in \Omega$ be arbitrary. Theorem 17 gives the representation

$$f(z) = (z - w)^{\nu_w(f)} u(z) \tag{28}$$

and hence

$$\frac{\partial}{\partial z} \log f(z) = \frac{\nu_w(f)}{z - w} + \frac{\partial}{\partial z} \log u(z) \tag{29}$$

from which we see that w is a pole of $\frac{\partial}{\partial z} \log f$ if and only if it is a zero of f and

$$\operatorname*{Res}_{z=w} f(z) = \nu_w(f). \tag{30}$$

Theorem 22 now completes the proof.

An important application of Theorem 23 is the following theorem, proved by Gauss in his doctoral thesis [30].

THEOREM 24 (FUNDAMENTAL THEOREM OF ALGEBRA)
If p is a polynomial then

$$\sum_{\zeta \in \mathbb{C}} \nu_\zeta(p) = \deg p. \tag{31}$$

In other words p has precisely $\deg p$ *complex zeroes when counted with multiplicities.*

Proof. To prevent our formulae from becoming ugly, we write N for the degree of p. Let $r > 0$ be such that no zero of p has absolute value r and let C_r be the circle of radius r about 0, oriented counterclockwise. Theorem 23 shows that

$$\sum_{|\zeta| < r} \nu_\zeta(p) = \frac{1}{2\pi i} \int_{C_r} \frac{\partial}{\partial z} \log p(z) \, dz. \tag{32}$$

Writing

$$p(z) = \sum_{n=0}^{N} a_n z^n \tag{33}$$

we find

$$\sum_{|\zeta|<r} v_{\zeta}(p) = \frac{1}{2\pi} \int_0^{2\pi} f(r,\theta)\,d\theta. \tag{34}$$

where

$$f(r,\theta) = \frac{Na_N + (N-1)a_{N-1}r^{-1}e^{-i\theta} + \cdots + a_1 r^{1-N}e^{(1-N)i\theta}}{a_N + a_{N-1}r^{-1}e^{-i\theta} + \cdots + a_0 r^{-N}e^{-(N)i\theta}}. \tag{35}$$

$f(r,\theta)$ converges to N uniformly in θ as r tends to infinity, so the limit of the right hand side in (34) exists and equals $N = \deg p$. The same is therefore true of the left hand side and the theorem is proved.

Exercises

1. Prove that Γ is meromorphic on $\mathbf{C}$. Where are its poles? What are their orders and residues?

2. Prove that $\tan z$ is meromorphic on $\mathbf{C}$. Where are its poles? What are their orders and residues?

3. Prove that if f is analytic near w and g is meromorphic near w, then fg is meromorphic near w. Prove that

$$\operatorname*{Res}_{z=w} f(z)g(z) = f(w) \operatorname*{Res}_{z=w} g(z)$$

   if g has a simple pole at w. Does this equation still hold if the pole is not simple?

4. Prove an analogue of Theorem 10 for meromorphic functions.

5. Prove an analogue of Theorem 18 for meromorphic functions.

6. Suppose $f : \Omega_1 \to \Omega_2$ is analytic and bijective. Let C be a closed contractible contour of index 1 with respect to $f^{-1}(w)$. Evaluate

$$\frac{1}{2\pi i} \int_C \frac{z \frac{\partial f}{\partial z}(z)\,dz}{f(z) - w}.$$

7. Make another attempt at exercise 6 from Section 2.2.

2.7 Bernoulli Polynomials Revisited

We have two methods available for determining the coefficients of a convergent power series. By Taylor's theorem these can be computed by taking derivatives. On the other hand, using equation (9) of Section 2.4, we can evaluate the same coefficients by contour integration. The connection between theses two methods is supplied by the Cauchy integral formula in the form of equation (2) from Section 2.3. Which of these should we use? In fact it is best to use both and compare the results. This often produces nontrivial identities. In this section we will apply this idea to the generating function of the Bernoulli polynomials. This will allow us to compute their Fourier series and, in the process, the values of the zeta function at the even integers. As a byproduct we will find sum and product representations for the trigonometric functions. This will prove useful in Section 3.2.

This section also introduces an important technique, that of deforming the contour of integration through a series of poles. This idea will return in sections 2.8 and 3.2.

We last met the function

$$u(x, z) = \begin{cases} 1 & \text{if } z = 0, \\ \frac{ze^{zx}}{e^z - 1} & \text{if } z \neq 0 \end{cases} \tag{1}$$

in Section 1.7, where we defined the Bernoulli polynomials by

$$B_k(x) = \frac{\partial^k u}{\partial z^k}\Big|_{z=0}. \tag{2}$$

For any $x \in \mathbf{C}$ is is clear that $u(x, z)$ has simple poles at the points $z = 2\pi i n$ where n is a nonzero integer, and has no other poles. The residues at these poles are

$$\operatorname*{Res}_{z=2\pi in} u(x, z) = 2\pi i n e^{2\pi i n x}. \tag{3}$$

In particular $u(x, z)$ is analytic for $|z| < 2\pi$, so

$$u(x, z) = \sum_{k=0}^{\infty} \frac{B_k(x)}{k!} z^k. \tag{4}$$

by Theorem 16.

By equation (2) from Section 2.3 we can also write $B_k(x)$ as a contour integral,

$$B_k(x) = \frac{k!}{2\pi i} \int_C z^{-k-1} u(x, z) \, dz. \tag{5}$$

The requirements on the contour C are that it should be contractible in the region where u is analytic and that it should have index 1 with respect to 0. Let C_N be a counterclockwise-oriented square contour with sides of length $(4N+2)\pi$ parallel to the axes, and center 0. Clearly C_0 satisfies the requirements, so

$$B_k(x) = \frac{k!}{2\pi i} \int_{C_0} z^{-k-1} u(x, z)\, dz. \tag{6}$$

We will break C_N up into its four constituent sides C_{left}, C_{right}, C_{upper}, and C_{lower}.

Next we consider the integral

$$\frac{k!}{2\pi i} \int_{C_N} z^{-k-1} u(x, z)\, dz$$

for $N > 0$. This has index 1 with respect to the poles at $z = 2\pi i n$ with $-N \le n \le N$, and index 0 with respect to the others. By Theorem 22 we know that

$$\frac{k!}{2\pi i} \int_{C_N} z^{-k-1} u(x, z)\, dz = \frac{k!}{2\pi i} \int_{C_0} z^{-k-1} u(x, z)\, dz \tag{7}$$

$$+ k! \sum_{\substack{-N \le n \le N \\ n \ne 0}} \operatorname*{Res}_{z=2\pi i n} z^{-k-1} u(x, z).$$

We have already evaluated the first integral, and the residues at the nonzero poles are easily evaluated using exercise 3 of Section 2.6,

$$\frac{k!}{2\pi i} \int_{C_N} z^{-k-1} u(x, z)\, dz = B_k(x, z) + \sum_{\substack{-N \le n \le N \\ n \ne 0}} (2\pi i n)^{-k} e^{2\pi i n x} \tag{8}$$

or, equivalently,

$$B_k(x, z) = -k! \sum_{\substack{-N \le n \le N \\ n \ne 0}} (2\pi i n)^{-k} e^{2\pi i n x} - \frac{k!}{2\pi i} \int_{C_N} z^{-k-1} u(x, z)\, dz. \tag{9}$$

Next we show that the integral on the right in equation (9) tends to zero. For this we must suppose that $k > 1$ and $0 \le x \le 1$. We will use the estimate (12) of Section 2.1. The time-consuming part of the calculation is finding estimates on

$$|u(x, z) z^{-k-1}| = \left| \frac{e^{xz}}{e^z - 1} \right| |z|^{-k}$$

on C_N. It is easy to show that this quantity is bounded uniformly on C_N for any particular N, but this is useless because we need bounds which are uniform in N.

On C_{right} we have

$$\left|\frac{e^{xz}}{e^z - 1}\right| \leq \frac{e^{(2N+1)\pi x}}{e^{(2N+1)\pi} - 1} \tag{10}$$

$$= \frac{e^{(2N+1)\pi}}{e^{(2N+1)\pi} - 1} e^{(2N+1)\pi(x-1)}.$$

Since $N \geq 0$ we have

$$\left|\frac{e^{xz}}{e^z - 1}\right| \leq \frac{e^\pi}{e^\pi - 1} e^{(2N+1)\pi(x-1)} \tag{11}$$

and, since $x \leq 1$,

$$\left|\frac{e^{xz}}{e^z - 1}\right| \leq \frac{e^\pi}{e^\pi - 1}. \tag{12}$$

Similarly, on C_{left} we have

$$\left|\frac{e^{xz}}{e^z - 1}\right| \leq \frac{e^{-(2N+1)\pi x}}{1 - e^{-(2N+1)\pi}} \tag{13}$$

$$= \frac{e^{(2N+1)\pi}}{e^{(2N+1)\pi} - 1} e^{-(2N+1)\pi x}.$$

Since $N \geq 0$ we have

$$\left|\frac{e^{xz}}{e^z - 1}\right| \leq \frac{e^\pi}{e^\pi - 1} e^{-(2N+1)\pi x} \tag{14}$$

and, since $x \geq 0$,

$$\left|\frac{e^{xz}}{e^z - 1}\right| \leq \frac{e^\pi}{e^\pi - 1}. \tag{15}$$

On the C_{upper} and C_{lower} we have

$$\left|\frac{e^{xz}}{e^z - 1}\right| = \frac{e^{x\,\text{Re}\,z}}{e^{\text{Re}\,z} + 1}, \tag{16}$$

or, since $0 \leq x \leq 1$,

$$\left|\frac{e^{xz}}{e^z - 1}\right| \leq 1. \tag{17}$$

In all cases

$$\left|\frac{e^{xz}}{e^z - 1}\right| \leq \frac{e^\pi}{e^\pi - 1}. \tag{18}$$

Since

$$|z|^{-k} \le (2N + 1)^{-k} \pi^{-k} \tag{19}$$

we have the estimate

$$|u(x, z)z^{-k-1}| \le \frac{\pi^{-k} e^\pi}{e^\pi - 1}(2N + 1)^{-k} \tag{20}$$

uniformly for $z \in C_N$. The length of the contour is $8(2N + 1)\pi$, so the estimate (12) of Section 2.1 shows that

$$|\frac{k!}{2\pi i} \int_{C_N} z^{-k-1}u(x, z)\, dz| \le \frac{4\pi^{-k} e^\pi}{e^\pi - 1}(2N + 1)^{1-k}. \tag{21}$$

Since we have assumed that $k > 1$ the right hand side tends to zero as N tends to infinity, so we find

$$B_k(x) = -k! \lim_{N\to\infty} \sum_{\substack{-N \le n \le N \\ n \ne 0}} (2\pi i n)^{-k} e^{2\pi i n x}. \tag{22}$$

Grouping the terms with opposite signs, we get

$$B_k(x) = -k! \lim_{N\to\infty} \sum_{n=1}^{N} (2\pi i n)^{-k}[e^{2\pi i n x} + (-1)^k e^{2\pi i n x}]. \tag{23}$$

or

$$B_k(x) = \begin{cases} (-1)^{(k+2)/2} k! 2(2\pi)^{-k} \sum_{n=1}^{\infty} \cos(2\pi n x) & \text{if } k \text{ is even,} \\ (-1)^{(k+1)/2} k! 2(2\pi)^{-k} \sum_{n=1}^{\infty} \sin(2\pi n x) & \text{if } k \text{ is odd.} \end{cases} \tag{24}$$

The case $x = 0$ of equation (24),

$$B_k = \begin{cases} (-1)^{(k+2)/2} k! 2(2\pi)^{-k} \sum_{n=1}^{\infty} \zeta(k) & \text{if } k \text{ is even,} \\ 0 & \text{if } k \text{ is odd,} \end{cases} \tag{25}$$

is particularly interesting. It allows us to compute the values of ζ at the even integers,

$$\zeta(k) = (-1)^{(k+2)/2} \frac{1}{2} \frac{B_k}{k!}(2\pi)^k. \tag{26}$$

Unfortunately no similar trick is known for the odd integers.

We have proved equation (24) only for $k > 1$ and $0 \le x \le 1$. In fact it remains true for $k = 1$, provided that $0 < x < 1$. This will be proved by contour integration as well, but the estimates on the integrals are more delicate. On C_{right} and C_{left} we use the estimates (11) and (14), respectively. From the estimate (12) of Section 2.1 it follows that

$$|\int_{C_{\text{left}}} u(x, z)z^{-2}\, dz| \leq \frac{2e^{\pi}}{e^{\pi} - 1} e^{(2N+1)\pi(x-1)} \tag{27}$$

and

$$|\int_{C_{\text{right}}} u(x, z)z^{-2}\, dz| \leq \frac{2e^{\pi}}{e^{\pi} - 1} e^{-(2N+1)\pi x}. \tag{28}$$

In both cases, since $0 < x < 1$ the right hand sides tend to zero as N tends to infinity. The integrals along C_{upper} and C_{lower} present greater difficulties because $\frac{e^{xz}}{e^z-1}$ is not $o(1)$ for N large. We therefore cannot use the estimate (12) of Section 2.1. Instead we proceed as follows:

$$\begin{aligned}
|\int_{C_{\text{lower}}} u(x, z)z^{-2}\, dz| &= |\int_{C_{\text{lower}}} \frac{e^{xz}}{e^z - 1} z^{-1}\, dz| \tag{29} \\[2mm]
&\leq \max |z^{-1}| \int_{C_{\text{lower}}} |\frac{e^{xz}}{e^z - 1}|\, |dz| \\[2mm]
&= \max |z^{-1}| \int_{C_{\text{lower}}} \frac{e^{x\,\text{Re}\,z}}{e^{\text{Re}\,z} + 1}\, |dz| \\[2mm]
&= \max |z^{-1}| \int_{-(2N+1)\pi}^{(2N+1)\pi} \frac{e^{x\xi}}{e^{\xi} + 1}\, d\xi \\[2mm]
&= \max |z^{-1}| \int_{-\infty}^{\infty} \frac{e^{x\xi}}{e^{\xi} + 1}\, d\xi \\[2mm]
&= (2N + 1)^{-1}\pi^{-1} \int_{-\infty}^{\infty} \frac{e^{x\xi}}{e^{\xi} + 1}\, d\xi.
\end{aligned}$$

The integral on the right is convergent for $0 < x < 1$ and is independent of N. The factor $(2N + 1)^{-1}$ thus ensures that the right hand side tends to zero as N tends to infinity. The argument for C_{upper} is identical. Thus we see that the integral along C_N tends to 0 as N tends to infinity, and therefore that (23) and (24) hold. It is quite easy to see that these equations do not hold for $k = 1$ and $x = 0$ or $x = 1$. Indeed $B_1(1) = \frac{1}{2}$ and $B_1(0) = -\frac{1}{2}$ while the right hand side of equation (24) would give zero in both cases.

The power series (4) is not the only useful expansion for $u(x, z)$. Replacing $B_k(x)$ for $k > 0$ by (23) we find

$$u(x, z) = 1 + a(x)z - \sum_{k=1}^{\infty} \lim_{N \to \infty} \sum_{\substack{-N \leq n \leq N \\ n \neq 0}} (\frac{z}{2\pi i n})^k e^{2\pi i n x}. \tag{30}$$

We have inserted an extra term $a(x)z$ with

$$
a(x) = \begin{cases} -\frac{1}{2} & \text{if } x = 0 \\ 0 & \text{if } 0 < x < 1 \\ \frac{1}{2} & \text{if } x = 1 \end{cases} \tag{31}
$$

to make (30) correct even in the exceptional cases $x = 0$ and $x = 1$. We would like to switch the order of summation, but the double sum is not absolutely convergent. The way around this problem is to split off the $k = 1$ term, switch the order of the sums, and then restore the $k = 1$ term.

$$
\begin{aligned}
u(x, z) &= 1 + a(x)z - \lim_{N \to \infty} \sum_{\substack{-N \le n \le N \\ n \ne 0}} \left(\frac{z}{2\pi i n} \right) e^{2\pi i n x} \tag{32} \\
&\quad - \sum_{k=2}^{\infty} \lim_{N \to \infty} \sum_{\substack{-N \le n \le N \\ n \ne 0}} \left(\frac{z}{2\pi i n} \right)^{k} e^{2\pi i n x} \\
&= 1 + a(x)z - \lim_{N \to \infty} \sum_{\substack{-N \le n \le N \\ n \ne 0}} \left(\frac{z}{2\pi i n} \right) e^{2\pi i n x} \\
&\quad - \lim_{N \to \infty} \sum_{\substack{-N \le n \le N \\ n \ne 0}} \sum_{k=2}^{\infty} \left(\frac{z}{2\pi i n} \right)^{k} e^{2\pi i n x} \\
&= 1 + a(x)z - \lim_{N \to \infty} \sum_{\substack{-N \le n \le N \\ n \ne 0}} \sum_{k=1}^{\infty} \left(\frac{z}{2\pi i n} \right)^{k} e^{2\pi i n x}.
\end{aligned}
$$

The interchange of the two infinite sums is justified by the absolute convergence of the double sum. Otherwise we have only switched to order of limits and *finite* sums. As it turns out, this precaution was unnecessary. If we had blindly moved the sum over k past the limit and the sum over n then we would have obtained the correct answer. In mathematics, as in life, virtue is not always rewarded, nor vice always punished. In any case we obtain the equation

$$
u(x, z) = 1 + a(x)z - \lim_{N \to \infty} \sum_{\substack{-N \le n \le N \\ n \ne 0}} \sum_{k=1}^{\infty} \left(\frac{z}{2\pi i n} \right)^{k} e^{2\pi i n x}. \tag{33}
$$

The inner sum is, of course the sum of a geometric series and can be evaluated immediately.

$$
u(x, z) = 1 + a(x)z + \lim_{N \to \infty} \sum_{\substack{-N \le n \le N \\ n \ne 0}} \frac{z}{z - 2\pi i n} e^{2\pi i n x} \tag{34}
$$

or, more simply,

$$u(x, z) = a(x)z + \lim_{N \to \infty} \sum_{-N \le n \le N} \frac{z}{z - 2\pi i n} e^{2\pi i n x}. \tag{35}$$

Recalling the equation (1) which defined $u(x, z)$, we see that

$$\frac{e^{zx}}{e^z - 1} = a(x) + \lim_{N \to \infty} \sum_{-N \le n \le N} \frac{e^{2\pi i n x}}{z - 2\pi i n}. \tag{36}$$

Because we began from the expansion (4) we have only proved (36) for $|z| < 2\pi$. Since both sides are analytic for $z \ne 2\pi i n$, we can use Theorem 18 to remove that restriction. The most interesting cases of equation (36) occur when $x = 0$, $x = \frac{1}{2}$, or $x = 1$, These give

$$\frac{e^{-z/2}}{e^{z/2} - e^{-z/2}} = -\frac{1}{2} + \lim_{N \to \infty} \sum_{-N \le n \le N} \frac{1}{z - 2\pi i n}, \tag{37}$$

$$\frac{1}{e^{z/2} - e^{-z/2}} = \lim_{N \to \infty} \sum_{-N \le n \le N} \frac{(-1)^n}{z - 2\pi i n}, \tag{38}$$

and

$$\frac{e^{z/2}}{e^{z/2} - e^{-z/2}} = \frac{1}{2} + \lim_{N \to \infty} \sum_{-N \le n \le N} \frac{1}{z - 2\pi i n}. \tag{39}$$

Adding (37) and (39) we find

$$\frac{e^{z/2} + e^{z/2}}{e^{z/2} - e^{-z/2}} = 2 \lim_{N \to \infty} \sum_{-N \le n \le N} \frac{1}{z - 2\pi i n}. \tag{40}$$

Setting $z = \pi i x$ we get the expansions

$$\pi \csc \pi x = \lim_{N \to \infty} \sum_{-N \le n \le N} \frac{1}{x - n}. \tag{41}$$

and

$$\pi \cot \pi x = \lim_{N \to \infty} \sum_{-N \le n \le N} \frac{(-1)^n}{x - n} \tag{42}$$

from equations (38) and (40). These expansions are due to Euler and are given, with an entirely different proof, in [26]. Integrating equation (42) we see that

$$\log \sin \pi x = \log x + \sum_{n=1}^{\infty} (\log(n^2 - x^2) - \log(n^2)) + c, \qquad (43)$$

where c is a constant of integration. Exponentiating we find

$$\sin \pi x = e^c x \prod_{n=1}^{\infty} \left(1 - \frac{x^2}{n^2}\right). \qquad (44)$$

The constant e^c can be evaluated easily by examining the derivatives of both sides at $x = 0$,

$$\sin \pi x = \pi x \prod_{n=1}^{\infty} \left(1 - \frac{x^2}{n^2}\right). \qquad (45)$$

This representation is also due to Euler. The formula

$$\Gamma(x)\Gamma(1-x) = \pi \csc \pi x, \qquad (46)$$

which was so difficult to prove in Section 1.1 follows immediately from (45) and the Weierstrass product representation.

Exercises

1. What is $\sin \frac{\pi}{2}$? Use the product representation for the sine to obtain the product formula

$$\frac{\pi}{2} = \frac{2 \cdot 2 \cdot 4 \cdot 4 \cdot 6 \cdot 6 \cdots}{1 \cdot 3 \cdot 3 \cdot 5 \cdot 5 \cdot 7 \cdots},$$

   first given in Wallis' book [69].

2. Evaluate $\zeta(2k)$ for $k = 1, 2, 3, 4, 5$.

3. Find the Laurent expansion of Γ'/Γ.

4. Prove that

$$\lim_{N \to \infty} \sum_{-N \leq n \leq N} (x + n)^{-1} = -\pi i - 2\pi i \sum_{l=1}^{\infty} q^{-2l}$$

   where $q = e^{\pi i x}$.

5. Prove that

$$\sum_{n=-\infty}^{\infty} (x+n)^{-k} = \frac{(-2\pi i)^k}{(k-1)!} \sum_{l=1}^{\infty} l^k q^{-2l}$$

for $k > 1$ where $q = e^{\pi i x}$.

6. Show that

$$\lim_{k\to\infty} \frac{B_{2k}}{(2\pi)^{2k}(2k)!} = \frac{1}{2}.$$

7. Show that the series

$$\sum_{k=1}^{\infty} (-1)^k \frac{B_k}{k!}(s)_{k-1}$$

from exercise 3 of Section 1.8 diverges for all s.

8. Find product representations for the cosine and tangent.

9. Prove that

$$\max_{0 \le x \le 1} |B_m(x)| = |B_m|$$

for m even.

10. In Section 3.2 we will meet the functions

$$E_{2k}(\tau) = \sum (m\tau + n)^{-2k}$$

where $k > 1$ is an integer and the sum ranges over all pairs (m, n) of integers. Prove the identity

$$E_{2k}(\tau) = (-1)^{k+1} \frac{(2\pi)^{2k}}{(2k-1)!} [\frac{B_{2k}}{2k} - 2\sum_{n=1}^{\infty} \sigma_{2k-1}(n)q^{2n}],$$

where

$$\sigma_{2k-1}(n) = \sum_{d|n} d^{2k-1}.$$

The sum runs over positive divisors of n.

2.8   Mellin-Barnes Integrals I

In the preceding section we introduced the technique of deforming the contour of integration out to infinity. In this section we will apply this idea to the study of the hypergeometric and Whittaker functions. In chapter 1 we were only able to determine the expansions of the Whittaker function $J(p, q, s)$ about $s = \infty$ and of the hypergeometric function $F(a, b; c; x)$ about $x = 0$. In this section we will show how to determine the expansion of the Whittaker function about $s = 0$ and of the hypergeometric function about $x = 1$ and $x = \infty$. The proofs involve some rather complicated estimates on contour integrals, and are deferred to the next section. Here we will only sketch the ideas of the proofs.

The main idea of this section is due to Barnes [2]. Consider the function

$$k(t, z) = -\Gamma(-z)t^z. \tag{1}$$

As a function of z, with t held fixed, k has simple poles at the points $z = n$ for nonnegative integral n. To see this we apply the functional equation

$$\Gamma(z + m) = (z)_m \Gamma(z) \tag{2}$$

with $m = n + 1$ to derive

$$k(t, z) = (z - n)^{-1} \frac{\Gamma(n + 1 - z)t^z}{(z)_n}. \tag{3}$$

The second factor is analytic near $z = n$. Since

$$\lim_{z \to -n} \frac{\Gamma(n + 1 - z)t^z}{(z)_n} = \frac{\Gamma(1)t^n}{(-n)_n} = (-1)^n \frac{t^n}{(1)_n} \tag{4}$$

we see that $z = n$ is a simple pole with residue

$$\operatorname*{Res}_{z=-n} k(t, z) = (-1)^n \frac{t^n}{(1)_n}. \tag{5}$$

To evaluate $(-n)_n$ we have used the identity

$$(\alpha - n)_n = (-1)^n (1 - \alpha)_n, \tag{6}$$

which is proved by making the substitution $k = n - j - 1$ in the product

$$(\alpha - n)_n = \prod_{j=0}^{n-1} (a + j) \tag{7}$$

defining the Pochhammer symbol. We will need the identity (6) a few more times in this section. There are no other poles since both $\Gamma(-z)$ and t^z are analytic on the complement of the nonnegative integers. If g is some meromorphic function whose poles are disjoint from those of k, then

$$\operatorname*{Res}_{z=-n} k(t, z)g(z) = (-1)^n \frac{g(n)}{(1)_n} t^n. \tag{8}$$

If there were a closed contour with index 1 with respect to the poles of k and index 0 with respect to the poles of g then the residue theorem for meromorphic functions, Theorem 22, would imply the power series expansion

$$f(t) = \sum_{n=0}^{\infty} (-1)^n \frac{g(n)}{(1)_n} t^n \tag{9}$$

for the function

$$f(t) = \frac{1}{2\pi i} \int_C k(t, z)g(z)\, dz.$$

There is no such contour. We can, however, construct a contour C_N with index 1 with respect to the points $0, 1, \ldots, N$ and index 0 with respect to all other poles of kg. If we define

$$f(t) = \lim_{N \to \infty} \frac{1}{2\pi i} \int_{C_N} k(t, z)g(z)\, dz, \tag{10}$$

then equation (9) still follows from Theorem 22. Of course in any particular example this limit may diverge for some or even all values of t. We must expect that we will have to do some work. At a minimum we will need estimates on the Gamma function in the complex domain.

We now consider three examples to see what type of theorem we may hope to prove by these methods. A particularly simple example is

$$g(z) = \frac{\Gamma(z + a)}{\Gamma(a)}. \tag{11}$$

g has simple poles at the points $z = -a - n$. Our standing assumption that the poles of g and k are disjoint excludes the case where a is a nonpositive integer. Equation (9) gives

$$f(t) = \sum_{n=0}^{\infty} (-1)^n \frac{(a)_n}{(1)_n} t^n = (1 + t)^{-a}. \tag{12}$$

Consider now a sequence of contours C_N' with index 1 with respect to the poles $-a, -a - 1, \ldots, -a - N$ of g and index 0 with respect to all other poles of kg.

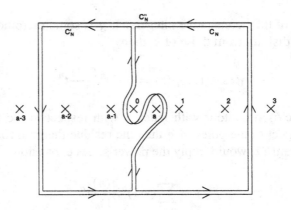

Figure 2.1: Contours for $(1-t)^{-a}$.

The residues at these poles are

$$\operatorname*{Res}_{z=-a-n} k(t,z)g(z) = \Gamma(a+n)t^{-a-n} \operatorname*{Res}_{z=-a-n} \frac{\Gamma(z+a)}{\Gamma(a)} \tag{13}$$

$$= (-1)^{n+1}\frac{(a)_n}{(1)_n}t^{-a-n}.$$

Theorem 22 shows that

$$-\lim_{N\to\infty}\frac{1}{2\pi i}\int_{C_N'} k(t,z)g(z)\,dz = \sum_{n=0}^{\infty}(-1)^{n+1}\frac{(a)_n}{(1)_n}t^{-a-n} = t^{-a}(1+t^{-1})^{-a}. \tag{14}$$

Raising the elementary identity

$$1+t = t(1+t^{-1}) \tag{15}$$

to the power $-a$, we find[2]

$$(1+t)^{-a} = t^{-a}(1+t^{-1})^{-a}. \tag{16}$$

We can express this identity in terms of contour integrals. Suppose the contour C_N'' is formed by joining the contours C_N and C_N'. The case $a = \frac{1}{2}$, $N = 2$ is illustrated in the diagram. We conclude from (16) that

---

[2]This is not quite as straightforward as it seems, as Exercise 9 of Section 2.2 shows. The reader should pause for a moment to understand why the pathology alluded to in that exercise does not occur here.

$$\lim_{N \to \infty} \frac{1}{2\pi i} \int_{C_N''} k(t, z) g(z) \, dz = 0 \tag{17}$$

wherever both the series (12) and (14) converge. This, however, happens only rarely. If $\text{Re}\, a > 1$ then there is no t for which both series converge. Even in the most favorable case,[3] $\text{Re}\, a < 0$, we have convergence of both series only for $|t| = 1$.

The preceding argument is somewhat reversible. If we can prove (17) for some values of a and t, then we can conclude that (16) holds for those same values, without the use of the identity (15). As mentioned above, the values for which this can be accomplished are rather few. Suppose, however, that we can prove this for $\text{Re}\, a < 0$ and $|t| = 1$. Then we can extend the identity (16) to all a and to all t off the negative real axis by applying Theorem 18 twice, once in the variable a, and once in the variable t. We defer for a moment the question of why we would want to give such a complicated proof of such a simple identity and proceed to the next example.

If the technique sketched above applied only to the example (11) and produced only the identity (16) then it would be of very little value. Suppose, however, that we apply the same ideas to the function

$$g(z) = \frac{\Gamma(z + a)\Gamma(z + b)\Gamma(c)}{\Gamma(a)\Gamma(b)\Gamma(z + c)}. \tag{18}$$

Equation (9) then gives

$$f(t) = \sum_{n=0}^{\infty} \frac{(a)_n (b)_n}{(1)_n (c)_n} t^n = F(a, b; c; -t). \tag{19}$$

The middle term in this equation is the expansion about $t = 0$. We may suspect, by analogy with the previous example, that minus the sum of the residues at the poles of g should give the expansion about $t = \infty$. This is in fact true. There are two series of poles of g, those at $z = -a - n$ and those at $z = -b - n$. As before we assume that none of these coincide with poles of k, in other words, that neither a nor b is a nonnegative integer. We will also assume that none of the poles from one series coincides with one from the other series, in other words, that $a - b$ is

---

[3] Recall that we have already excluded the case where a is a nonpositive integer.

not an integer. The residues at $z = -a - n$ are evaluated as follows:

$$
\operatorname*{Res}_{z=-a-n} k(t, z)g(t, z) = - \operatorname*{Res}_{z=-a-n} \frac{\Gamma(-z)\Gamma(z + a)\Gamma(z + b)\Gamma(c)}{\Gamma(a)\Gamma(b)\Gamma(z + c)} t^z \tag{20}
$$

$$
= -\frac{\Gamma(a + n)\Gamma(b - a - n)\Gamma(c)}{\Gamma(a)\Gamma(b)\Gamma(c - a - n)} t^{-a-n} \operatorname*{Res}_{z=-a-n} \Gamma(z + a)
$$

$$
= \frac{(-1)^{n+1}}{(1)_n} \frac{\Gamma(a + n)\Gamma(b - a - n)\Gamma(c)}{\Gamma(a)\Gamma(b)\Gamma(c - a - n)} t^{-a-n}
$$

$$
= (-1)^{n+1} \frac{\Gamma(c)\Gamma(b - a)}{\Gamma(b)\Gamma(c - a)} \frac{(a)_n(c - a - n)_n}{(1)_n(b - a - n)_n} t^{-a-n}
$$

$$
= (-1)^{n+1} \frac{\Gamma(c)\Gamma(b - a)}{\Gamma(b)\Gamma(c - a)} \frac{(a)_n(1 + a - c)_n}{(1)_n(1 + a - b)_n} t^{-a-n}.
$$

In the second to last line we used equation (2). In the last line we used the identity (6). We need not duplicate this effort to compute the residues at the poles $z = -b - n$; it suffices to note that g is symmetric in the variables a and b, so that

$$
\operatorname*{Res}_{z=-b-n} k(t, z)g(t, z) = (-1)^{n+1} \frac{\Gamma(c)\Gamma(a - b)}{\Gamma(a)\Gamma(c - b)} \frac{(b)_n(1 + b - c)_n}{(1)_n(1 + b - a)_n} t^{-b-n}. \tag{21}
$$

If we choose a series of contours C_N' which have index 1 with respect to $-a, -a - 1, \ldots, -a - N$ and $-b, -b - 1, \ldots, -b - N$, and index 0 with respect to the other poles of $k(t, z)g(z)$, then Theorem 22 shows that

$$
- \lim_{N \to \infty} \int_{C_N'} k(t, z)g(z)
$$

is equal to

$$
\frac{\Gamma(c)\Gamma(b - a)}{\Gamma(b)\Gamma(c - a)} t^{-a} F(a, 1 + a - c; 1 + a - b; -t^{-1})
$$

$$
+ \frac{\Gamma(c)\Gamma(a - b)}{\Gamma(a)\Gamma(c - b)} t^{-b} F(b, 1 + b - c; 1 + b - a; -t^{-1}).
$$

This, according to the conjecture stated above, should be the expansion about $t = \infty$ of $F(a, b; c; -t)$. The following theorem, which we will prove in the next section, confirms that conjecture.

THEOREM 25 (EXPANSION OF $F(a, b; c; x)$ ABOUT $x = \infty$)
If $a - b$ is not an integer, c is not a nonpositive integer, and $x \in \Omega$, where

$$
\Omega = \{x \in \Omega: 0 < \operatorname{Re} x < 1 \text{ or } \operatorname{Im} x \neq 0\}, \tag{22}
$$

then

$$F(a, b; c; x) = \frac{\Gamma(c)\Gamma(b-a)}{\Gamma(b)\Gamma(c-a)}(-x)^{-a}F(a, 1+a-c; 1+a-b; x^{-1}) \quad (23)$$
$$+ \frac{\Gamma(c)\Gamma(a-b)}{\Gamma(a)\Gamma(c-b)}(-x)^{-b}F(b, 1+b-c; 1+b-a; x^{-1}).$$

Theorem 25 provides a link between the hypergeometric function at x and at x^{-1}. This complements the three relations of Euler given as equation (38) in Section 1.6. Those related the hypergeometric function at x and at $\frac{x}{x+1}$. Let functions σ and τ be defined by

$$\sigma(x) = x^{-1} \quad (24)$$
$$\tau(x) = \frac{x}{x-1}.$$

By using equations (38) of Section 1.6 and equation (23) above, we can express $F(a, b; c; x)$ in terms of hypergeometric functions evaluated at $\alpha(x)$ where α is any function formed by composing the functions σ and τ any number of times in any order. This is, however, less impressive than it sounds. There are only six such functions. These are

$$\alpha_1(x) = x, \quad (25)$$
$$\alpha_2(x) = \sigma(x) = x^{-1},$$
$$\alpha_3(x) = \tau(x) = \frac{x}{x-1},$$
$$\alpha_4(x) = (\sigma \circ \tau)(x) = \frac{x-1}{x},$$
$$\alpha_5(x) = (\tau \circ \sigma)(x) = \frac{1}{1-x},$$
$$\alpha_6(x) = (\sigma \circ \tau \circ \sigma) = 1 - x.$$

It is now a matter of simple algebra to compile a list of the relations between the hypergeometric functions at x and $\alpha_j(x)$ for $j = 1, \ldots, 6$. Such a table may be found in many places, including [75], [22], and [35]. It is not, in fact, terribly useful. In order to use Theorem 25, one is forced to make the assumption that none of the numbers c, $a - b$, and $c - a - b$ is an integer. Unfortunately, in most examples which occur in practice, some or even all of these numbers are integers. There are ways of coping with this problem, which will be discussed briefly at the end of this section.

One could give a proof of Theorem 25 along the same lines as that suggested in the previous example, fusing the two contours C_N and C'_N into one contour C''_N

and then showing that

$$\lim_{N \to \infty} \frac{1}{2\pi i} \int_{C_N''} k(t, z) g(z) \, dz = 0 \tag{26}$$

for appropriate a, b, c, and t. In particular one can prove equation (26), and hence (22), under the hypotheses

$$\text{Re}(a + b - c) < 0 \tag{27}$$

and

$$|t| = 1, \tag{28}$$

and then use Theorem 18 to remove these restrictions. The method we will actually use is slightly different.

Before proceeding to the proofs we consider one more example. Taking

$$g(z) = \frac{\Gamma(z + p)\Gamma(z + p + q)}{\Gamma(p)\Gamma(p + q)} \tag{29}$$

we see that (9) gives

$$f(t) = \sum_{n=0}^{\infty} (-1)^n \frac{(p)_n (p + q)_n}{(1)_n} t^n. \tag{30}$$

The series is divergent, so the limit of the contour integrals cannot be convergent. This divergent series, however, is one which we have met before. It is the asymptotic expansion about $t = 0$ of the function $t^{-p} J(p, q, t^{-1})$ from Section 1.5. It is not unreasonable to guess that, as in the preceding two examples, minus the sum of the residues at the poles of g should give the expansion about $t = \infty$. There are two series of poles, those at $z = -p - n$ and those at $z = -p - q - n$. As in the previous example, we suppose that these two series are disjoint both from each other and from the poles of k. The residues are evaluated by the same method we used in the previous example. The result of this calculation is

$$\operatorname*{Res}_{z=-p-n} k(t, z) g(z) = -\frac{\Gamma(q)}{\Gamma(p + q)} \frac{(p)_n}{(1)_n (1 - q)_n} t^{-p-n} \tag{31}$$

$$\operatorname*{Res}_{z=-p-q-n} k(t, z) g(z) = -\frac{\Gamma(-q)}{\Gamma(p)} \frac{(p + q)_n}{(1)_n (1 + q)_n} t^{-p-q-n}.$$

Our conjecture, to be proved in the next section, is as follows.

THEOREM 26 (EXPANSION OF $J(p, q, s)$ ABOUT $s = 0$)
For q not an integer and s off the negative real axis

$$J(p, q, s) = \frac{\Gamma(q)}{\Gamma(p+q)} \sum_{n=0}^{\infty} \frac{(p)_n}{(1)_n(1-q)_n} s^n + \frac{\Gamma(-q)}{\Gamma(p)} \frac{(p+q)_n}{(1)_n(1+q)_n} s^{q+n}. \quad (32)$$

The method sketched for the previous two examples is powerless here because there are no p, q, s such that the expansions about $s = 0$ and $s = \infty$ are both convergent. Theorem 18 is powerful, but it cannot extend an identity from the empty set! We therefore need to change our approach. The proof given in the next section has four steps:

1. Choose an appropriate (infinite) contour C and show that the integral

$$\int_C k(t, z) g(z) \, dz$$

converges locally uniformly in p, q, and t.

2. Prove that

$$\int_C k(t, z) g(z) \, dz = \int_{C_N} k(t, z) g(z) \, dz + O(t^{\alpha+n}) \quad (33)$$

for $|t|$ small, with some α depending on p and q.

3. Prove that

$$\lim_{N \to \infty} \int_{C_N'} k(t, z) g(z) \, dz = - \int_C k(t, z) g(z) \, dz. \quad (34)$$

4. Prove that

$$J(p, q, s) = \frac{1}{2\pi i} \int_C k(t, z) g(z) \, dz. \quad (35)$$

Step 2 is irrelevant to the proof of Theorem 26, but it allows us to extend the asymptotic expansion of $J(p, q, s)$ beyond the region where the original integral representation converges. Step 4 is needed because analytic functions are not uniquely determined by their asymptotic expansions. Without step 4 we could only conclude that there is some function with the same asymptotic expansion at $s = \infty$ as $J(p, q, s)$ and the expansion (32) at $s = 0$; we would have no way of knowing that this function is really $J(p, q, s)$.

The proof we give for Theorem 25 follows similar lines, except that no analogue of step 4 is needed since a convergent power series expansion, unlike an asymptotic expansion, determines a function uniquely. The steps are:

1. Define an appropriate (infinite) contour C and show that the integral

$$\int_C k(t, z) g(z) \, dz$$

converges locally uniformly in p, q, and t.

2. Prove that

$$\lim_{N \to \infty} \int_{C_N} k(t, z) g(z) \, dz = \int_C k(t, z) g(z) \, dz. \tag{36}$$

for $|t|$ small, with some α depending on p and q.

3. Prove that

$$\lim_{N \to \infty} \int_{C'_N} k(t, z) g(z) \, dz = - \int_C k(t, z) g(z) \, dz. \tag{37}$$

Fortunately the integrals are all very similar, so we can, at the expense of a somewhat cumbersome notation, avoid much duplication of effort.

We have, in each example, imposed constraints on the parameters so that all the poles of $k(t, z)$ will be simple. This made the evaluation of the residues much easier, but it excludes many interesting examples. The formulae in the general case are quite complicated, as is their derivation. To illustrate what is involved we consider the case of the complete elliptic integral of the first kind, given by the formula

$$K(\kappa) = \frac{\pi}{2} F(\frac{1}{2}, \frac{1}{2}; 1; \kappa^2). \tag{38}$$

Theorem 25 is not directly applicable, but the formula

$$F(a, b; c; -t) = \sum \operatorname{Res} k(t, z) g(z), \tag{39}$$

where the sum is over the poles of g, remains valid. We are faced with the problem of computing the residue of

$$h(z) = \frac{1}{2} \frac{\Gamma(z + \frac{1}{2})^2 \Gamma(-z)}{\Gamma(z + 1)} (-\kappa^2)^z \tag{40}$$

at the points $z = -n - \frac{1}{2}$, which are poles of order 2. The actual calculation is rather complicated. The final answer is

$$\operatorname*{Res}_{s = -n - \frac{1}{2}} h(z) = (-\kappa^2)^{-1/2} \frac{(\frac{1}{2})_n^2}{(1)_n^2} \kappa^{-2n} [\frac{1}{2} \log(-\kappa^2) + 2 \log 2 + 2 \sum_{l=1}^{2n} \frac{(-1)^l}{l}].$$

$$\tag{41}$$

Usually one wants an expression in terms of κ rather than $-\kappa^2$. This is easily derived, but one must pay careful attention to the definitions of $\log x$ and x^y for complex x and nonintegral y. If we write

$$\kappa = re^{i\theta} \tag{42}$$

with r and θ real and $r > 0$, $-\pi < \theta < \pi$ then

$$-\kappa^2 = \begin{cases} r^2 e^{2i\theta + \pi i} & \text{if } -\pi < \theta < 0, \\ r^2 e^{2i\theta - \pi i} & \text{if } 0 < \theta < \pi. \end{cases} \tag{43}$$

In both cases we have written $-\kappa^2$ in the form

$$-\kappa^2 = \rho e^{i\phi} \tag{44}$$

with ρ and ϕ real and $\rho > 0$, $-\pi < \phi < \pi$. This makes it easy to compute the logarithm,

$$\log(-\kappa^2) = \begin{cases} 2\log r + 2i\theta + \pi i & \text{if } -\pi < \theta < 0, \\ 2\log r + 2i\theta - \pi i & \text{if } 0 < \theta < \pi, \end{cases} \tag{45}$$

or, in terms of $\log \kappa$,

$$\log(-\kappa^2) = \begin{cases} 2\log \kappa + \pi i & \text{if } -\pi < \theta < 0, \\ 2\log \kappa - \pi i & \text{if } 0 < \theta < \pi. \end{cases} \tag{46}$$

Since $(-\kappa^2)^{-1/2}$ is defined by

$$(-\kappa^2)^{-1/2} = \exp(-\frac{1}{2}\log(-\kappa^2)) \tag{47}$$

we find

$$(-\kappa^2)^{-1/2} = \begin{cases} -i\kappa & \text{if } -\pi < \theta < 0, \\ +i\kappa & \text{if } 0 < \theta < \pi. \end{cases} \tag{48}$$

Inserting these expressions into (41) and summing over n, we find

$$K(\kappa) = \pm i \sum_{n=0}^{\infty} \frac{(\frac{1}{2})_n^2}{(1)_n^2} \kappa^{-2n-1} [\log \kappa \pm \frac{1}{2}\pi i + 2\log 2 + 2\sum_{l=1}^{2n} \frac{(-1)^l}{l}], \tag{49}$$

where $\pm$ is to be interpreted as a $+$ if $\theta > 0$ and a $-$ if $\theta < 0$.

Exercises

1. Derive equation (41).

2. What should be the expansion about $x = \infty$ of

$$\sum_{n=0}^{\infty} \frac{(a)_n (b)_n (c)_n}{(1)_n (d)_n (e)_n} x^n?$$

3. What should be the expansion about $s = 0$ of $J(p, q, s)$ when q is an integer?

4. We have determined the behavior of $K(\kappa)$ under the transformation α_2. The behavior under α_1 is trivial. Determine the behavior under the other four transformations.

2.9   Mellin-Barnes Integrals II

The preceding section contained theorems but no proofs. This section will supply the missing proofs. This requires estimates on integrals of functions of the general form

$$h(x, z) = e^{xz} \prod_{j=1}^{m} \Gamma(s_j z - s_j a_j)^{p_j}, \qquad (1)$$

where x and the a_j's are complex, the p_j's are integers and the s_j's are ± 1. Since we do not assume that the a_j's are distinct there is no loss of generality in taking all the p_j's to be ± 1. Let m_{++} be the number of j's for which $p_j = +1, s_j = +1$, let m_{+-} be the number for which $p_j = +1, s_j = -1$, and so forth. We will see that the behavior of the integrals is controlled by the integers

$$k = m_{++} + m_{+-} - m_{-+} - m_{--} \qquad (2)$$

and

$$l = m_{++} - m_{+-} - m_{-+} + m_{--}. \qquad (3)$$

We have replaced the variable t of the preceding section with e^x for reasons which will be explained at the end of the section. We wish to make estimates which are locally uniform in x and in the a_j's so that Theorem 14 can be used to prove the analyticity of the integrals.

We will prove estimates for x, z in the region V defined by the following conditions.

1.

$$\mathrm{Re}\, x \le X_r, \qquad (4)$$

2.

$$-X_i \leq \operatorname{Im} x \leq X_i, \tag{5}$$

3.

$$|z| \geq 2\rho, \tag{6}$$

4.

$$-\frac{\pi}{2} \leq \arg z \leq \frac{\pi}{2}, \tag{7}$$

and

5. For all integers n and all $1 \leq j \leq m$

$$|z - a_j - n| \leq \frac{1}{4m}. \tag{8}$$

Here X_r is real, X_i positive real and ρ is a positive real number such that

$$\rho > 1 + \max_{1 \leq j \leq m} |a_j| \tag{9}$$

The region V depends only on the numbers m, X_i, X_r, ρ and on the a_j's. We now prove that

$$|h(x, z)| \leq C_4(m, \rho)|z|^{l \operatorname{Re} z + m\rho/2} e^{(X_r - l) \operatorname{Re} z - (l|\arg z| + (k-l)\frac{\pi}{2} - X_i)|\operatorname{Im} z|}. \tag{10}$$

for $(x, z) \in V$ with some C_4 depending only on m and ρ. We will give an explicit C_4, but we make no particular effort to get a sharp bound.

Our first task is to obtain estimates on $\Gamma(\zeta)$ for complex ζ. Already in Section (1.3) we saw that

$$\Gamma(\zeta) = (2\pi)^{1/2} \zeta^{\zeta - 1/2} e^{-\zeta + J(\zeta)} \tag{11}$$

with

$$0 \leq J(\zeta) \leq \frac{1}{12\zeta} \tag{12}$$

for positive real ζ. The proof of the inequality (12) given in that section cannot be extended to the complex ζ and we have already noted that the rigidity principle does not apply to inequalities. Fortunately, however, we can use most of Section 1.8 without change. We saw there that

$$J(\zeta) = \frac{1}{12\zeta} - \frac{1}{12} \int_0^\infty (\zeta + u)^{-2} \, du \tag{13}$$

We conclude that

$$|J(\zeta)| \le \frac{1}{12|\zeta|} + \frac{1}{12}\int_0^\infty |\zeta + u|^{-2}\,du = \frac{1}{12|\zeta|}(1 + \arg\zeta \csc\arg\zeta). \quad (14)$$

If

$$|z| \ge r \quad (15)$$

and

$$|\arg z| \le \theta \quad (16)$$

then

$$|J(\zeta)| \le \frac{1}{12r}(1 + \theta \csc\theta). \quad (17)$$

It is not difficult to extend this argument to show that Stirling's series gives an asymptotic expansion for $\Gamma(\zeta)$ in the region given by (15) and (16). Since this is not needed for the proof of (10) we will leave it as an exercise.

If $(x, z) \in V$ and $|a| + 1 \le \rho$ then

$$\zeta = z - a \quad (18)$$

satisfies (15) and (16) with $r = 1$ and $\theta = \frac{2\pi}{3}$. We conclude that

$$|J(z - a)| \le \frac{1}{12}(1 + \frac{4\sqrt{3}}{9}\pi). \quad (19)$$

It is not true in general that

$$\log(z - a) = \log z + \log(1 - az^{-1}) \quad (20)$$

The set where (20) is violated is shown in the diagram. Fortunately, however, it does not intersect V. We will need estimates on $\log(1 - az^{-1})$. Because of equation (6) we know that $|az^{-1}| \le \frac{1}{2}$. Therefore

$$|\log(1 - az^{-1})| = |\sum_{n=1}^\infty \frac{(-1)^n}{n}a^n z^{-1}| \le |az^{-1}|\sum_{n=1}^\infty \frac{2^{1-n}}{n} \le 2\log 2|az^{-1}| \quad (21)$$

and

$$|(z - a - \frac{1}{2})\log(1 - az^{-1})| \le 2\log 2|\frac{a(z - a - \frac{1}{2})}{z}| \le \frac{7}{4}\log 2|a| \le \frac{7}{2}\log 2\rho. \quad (22)$$

$$\log (z - a) = \log z + \log (1 - a z^{-1}) \text{ here,}$$

but not here.

Figure 2.2: Region of validity of equation 20.

We also have

$$|(a + \frac{1}{2}) \log z| \le \frac{\rho}{2} |\log z| \le \frac{\rho}{2} \operatorname{Re} \log z + \frac{\rho}{2} |\operatorname{Im} \log z| \le \frac{\rho}{2} \operatorname{Re} \log z + \frac{\pi \rho}{4} \tag{23}$$

Combining equations (20), (22), and (23) we find

$$|(z - a - \frac{1}{2}) \log(z + a) - z \log z - a| \le (\frac{7}{2} \log 2 + \frac{\pi}{4} + \frac{1}{2})\rho + \frac{\rho}{2} \operatorname{Re} \log z \tag{24}$$

Equations (11), (19) and (24) then show that

$$|\Gamma(z + a)| \le C_1(\rho)(2\pi)^{1/2}|z^{z+\rho/2}e^{-z}|, \tag{25}$$

and

$$|\Gamma(z + a)^{-1}| \le C_1(\rho)(2\pi)^{-1/2}|z^{-z+\rho/2}e^{z}|, \tag{26}$$

where

$$C_1(\rho) = \exp((\frac{7}{2} \log 2 + \frac{\pi \rho}{4} + \frac{1}{2})\rho + \frac{1}{12}(1 + \frac{4\sqrt{3}}{9}\pi)). \tag{27}$$

This takes care of the terms in (1) with $s_j = +1$. For those with $s_j = -1$ we use equation (46) of Section 2.7,

$$\Gamma(-z + a_j) = -\frac{\pi}{\Gamma(z - a + 1) \sin(\pi z - \pi a_j)} \tag{28}$$

From (8) we see that

$$\frac{1}{2}(1 - e^{-\pi/2m})e^{\pi|\operatorname{Im}(z - a_j)|} \le |\sin(\pi z - \pi a_j)| \le e^{\pi|\operatorname{Im}(z - a_j)|}, \tag{29}$$

while (6) and (9) show that

$$e^{-\pi\rho/2}e^{\pi|\operatorname{Im}z|} \le e^{\pi|\operatorname{Im}(z-a_j)|} \le e^{\pi\rho/2}e^{\pi|\operatorname{Im}z|}. \tag{30}$$

Using equation (25) and (26) to estimate $\Gamma(z-a_j+1)$ and equations (29) and (30) to estimate $|\sin(\pi z - \pi a_j)|$ we find

$$|\Gamma(-z+a_j)| \le C_2(m,\rho)|z|^{-z+\rho/2}e^z|e^{-\pi|\operatorname{Im}z|} \tag{31}$$

and

$$|\Gamma(-z+a_j)^{-1}| \le C_2(m,\rho)|z|^{+z+\rho/2}e^{-z}|e^{\pi|\operatorname{Im}z|} \tag{32}$$

where

$$C_2(m,\rho) = (2\pi)^{1/2}(1 - e^{-\pi/2m})^{-1}e^{\pi\rho/2}C_1(\rho) \tag{33}$$

and

$$C_3(m,\rho) = (\frac{\pi}{2})^{-1/2}e^{\pi\rho/2}C_1(\rho). \tag{34}$$

We now estimate each factor in the product (1) using one of (25), (26), (31) or (32), to obtain

$$|h(x,z)| \le C_4(m,\rho)|z|^{lz+m\rho/2}e^{xz-lz}|e^{(k-l)\frac{\pi}{2}|\operatorname{Im}z|} \tag{35}$$

where

$$C_4(m,\rho) = C_2(m,\rho)^m. \tag{36}$$

From the definition of complex powers we see that

$$|h(x,z)| \le C_4(m,\rho)|z|^{l\operatorname{Re}z+m\rho/2}e^{(\operatorname{Re}x-l)\operatorname{Re}z-(l\arg z+\operatorname{Im}x)\operatorname{Im}z+(k-l)\frac{\pi}{2}|\operatorname{Im}z|}. \tag{37}$$

$\arg z$ and $\operatorname{Im}z$ have the same sign so we may replace $\arg z \operatorname{Im}z$ by $|\arg z||\operatorname{Im}z|$. Doing so, and using (4) and (5) we obtain (10). We now have the estimates we need to prove the following theorem.

THEOREM 27 (ESTIMATES ON INTEGRALS OF BARNES' TYPE)
Let $h(x,z)$ be a function of the form (1) with $k > 0$ Let C be an (infinite) contour which follows the imaginary axis from $-i\infty$ to $-i\rho$, then proceeds to $+i\rho$ by any route which avoids the poles of h, and finally continues from $+i\rho$ to $+i\infty$ along the imaginary axis, and let $C_{\xi,\eta}$ be the (infinite) contour which starts at $-i\infty$ follows the imaginary axis to $-i\eta$, proceeds by straight line segments to $\xi - i\eta$, $\xi + i\eta$ and $+i\eta$, and then continues from to $+\infty$ along the imaginary axis.

1. *The integral*

$$f(x) = \frac{1}{2\pi i} \int_C h(x, z)\, dz \tag{38}$$

converges and defines an analytic function of x and $a_1, \ldots, a_m$ provided that

$$|\operatorname{Im} x| < \frac{k\pi}{2}. \tag{39}$$

2. *Let α be a real number such that*

$$|\alpha - \operatorname{Re} a_j - n| \geq \frac{1}{4m}. \tag{40}$$

If either

$$l < 0 \tag{41}$$

or

$$l = 0, \quad \operatorname{Re} x < 0 \tag{42}$$

then

$$\lim_{N \to \infty} \frac{1}{2\pi i} \int_{C_{N+\alpha, N+\alpha}} h(x, z)\, dz = 0. \tag{43}$$

3. *If $l > 0$ then, for any positive real λ and any positive $X_i \leq \frac{k\pi}{2}$, there are positive real ξ, η and K such that*

$$\left| \frac{1}{2\pi i} \int_{C_{\xi, \eta}} h(x, z)\, dz \right| \leq K e^{-\lambda \operatorname{Re} x} \tag{44}$$

for all x with $\operatorname{Re} x < 0$ and

$$|\operatorname{Im} x| \leq X_i. \tag{45}$$

Proof. We will need estimates on integrals over infinite contours. The inequality (12) of Section 2.1 is unsuitable for this purpose. We will estimate integrals of the form

$$\int_C f(\xi) \xi^{-2}\, d\xi.$$

If C is given by the path $\gamma : [a, b] \rightarrow \mathbf{C} - \{0\}$ then we define another path $\tilde{\gamma} : [a, b] \rightarrow \mathbf{C} - \{0\}$ by

$$\tilde{\gamma}(t) = 1/\gamma(t) \tag{46}$$

and compute

$$\int_C f(\xi)\xi^{-2} \, d\xi = \int_a^b \frac{f(\gamma(t)\gamma'(t) \, dt}{\gamma(t)^2} \tag{47}$$

$$= \int_a^b \tilde{f}(\tilde{\gamma}(t))\tilde{\gamma}'(t) \, dt$$

$$= \int_{\tilde{C}} \tilde{f}(\xi) \, d\xi$$

where

$$\tilde{f}(\xi) = f(1/\xi). \tag{48}$$

Applying the estimate (12) of Section 2.1 to the right hand side in (47) we find

$$\left| \int_C f(\xi)\xi^{-2} \, d\xi \right| \leq \max_{\xi \in C} |f(\xi)| \, \text{length} \, \tilde{C}. \tag{49}$$

We will need two special cases of this estimate. If C is a line segment then $\tilde{C}$ is an arc of a circle through the origin and

$$\text{length} \, \tilde{C} \leq \pi \max_{\xi \in \tilde{C}} |\xi| \qquad \bullet \tag{50}$$

so

$$\left| \int_C f(\xi)\xi^{-2} \, d\xi \right| \leq \pi \frac{\max_{\xi \in C} |f(\xi)|}{\min_{\xi \in C} |\xi|}. \tag{51}$$

If C is a polygonal contour with s sides then, applying (51) to each side,

$$\left| \int_C f(\xi)\xi^{-2} \, d\xi \right| \leq s\pi \frac{\max_{\xi \in C} |f(\xi)|}{\min_{\xi \in C} |\xi|}. \tag{52}$$

We begin with 1. We will prove the convergence of the improper integral by Cauchy's criterion. Choose an X_i in the range

$$|\operatorname{Im} x| < X_i < \frac{k\pi}{2}. \tag{53}$$

Let

$$T = \frac{1}{2} C_4(m, \rho)^{-1} \left(\frac{2 + m\rho/2}{\frac{k\pi}{2} - X_i} \right)^{-2-m\rho/2} \epsilon \qquad (54)$$

If $t_4 \geq t \geq t_3 \geq T$ then (10) gives the estimate

$$|h(x, it)| \leq C_4(m, \rho) y^{m\rho/2} e^{-(\frac{k\pi}{2} - X_i)t}. \qquad (55)$$

By equation (23) of the appendix

$$y^{m\rho/2} e^{-(\frac{k\pi}{2} - X_i)t} \leq M y^{-2} \qquad (56)$$

where

$$M = \left(\frac{2 + m\rho/2}{\frac{k\pi}{2} - X_i} \right)^{2 + m\rho/2}, \qquad (57)$$

and hence

$$|h(x, it)| \leq C_4(m, \rho) M t^{-2}. \qquad (58)$$

It follows that

$$\left| \int_{it_3}^{it_4} h(x, z) \, dz \right| \leq C_4(m, \rho) M (t_3^{-1} - t_4^{-1}) \leq C_4(m, \rho) M T^{-1} = \frac{\epsilon}{2}. \qquad (59)$$

Similarly, if $t_1 \leq t_2 \leq -T$ then

$$\left| \int_{it_1}^{it_2} h(x, z) \, dz \right| \leq \frac{\epsilon}{2} \qquad (60)$$

Combining these we find

$$\left| \int_{it_3}^{it_4} h(x, z) \, dz + \int_{it_1}^{it_2} h(x, z) \, dz \right| \leq \epsilon, \qquad (61)$$

so the integral converges by Cauchy's criterion. Any nearby $a_1, \ldots, a_m$ and x will satisfy (9) and (53) with the same ρ and X_i and thus will satisfy (61) with the same T. The convergence is therefore locally uniform in $a_1, \ldots, a_m$ and x. Analyticity follows from Theorem 14.

Next we prove 2. We write

$$l |\arg z| + (k - l)\frac{\pi}{2} = \frac{k\pi}{2} - l\left(\frac{\pi}{2} - |\arg z| \right) \qquad (62)$$

and hence, for $l \geq 0$,

$$l|\arg z| + (k - l)\frac{\pi}{2} \geq \frac{k\pi}{2} \tag{63}$$

We will consider first the case $l > 0$. Let

$$Z_1 = \exp(1 - l^{-1}\operatorname{Re} x - l^{-1}(\frac{k\pi}{2} - |\operatorname{Im} x|)) \tag{64}$$

$$M = \left(\frac{2 + m\rho/2}{\frac{k\pi}{2} - |\operatorname{Im} x|}\right)^{2+m\rho/2}, \tag{65}$$

and

$$Z_2 = \frac{5\pi C_4(m, \rho)M}{\epsilon}. \tag{66}$$

If $|z| > \max(Z_1, Z_2)$ then

$$|z|^{l \operatorname{Re} z} e^{(\operatorname{Re} x - l)\operatorname{Re} z} \leq e^{-(\frac{k\pi}{2} - |\operatorname{Im} x|)\operatorname{Re} z} \tag{67}$$

and therefore

$$
\begin{aligned}
|h(x, z)| &\leq C_4(m, \rho)e^{-(\frac{k\pi}{2} - |\operatorname{Im} x|)(\operatorname{Re} z + |\operatorname{Im} z|)} \\
&\leq C_4(m, \rho)e^{-(\frac{k\pi}{2} - |\operatorname{Im} x|)|z|}.
\end{aligned} \tag{68}
$$

It follows from equation (23) of the appendix that

$$|h(x, z)| \leq C_4(m, \rho)M|z|^{-2}. \tag{69}$$

The inequality (52) shows that

$$\left|\int_{C_{N+\alpha, N+\alpha}} h(x, z)\, dz\right| \leq 5\pi C_4(m, \rho)M \leq \epsilon(N + \alpha)^{-1} \tag{70}$$

if N is sufficiently large that

$$N + \alpha \geq \max(Z_1, Z_2). \tag{71}$$

The right hand side of (70) tends to zero as N tends to infinity, so 2 follows. The proof of 2 when $l = 0$ is similar, but simpler. From (63) we see that

$$|h(x, z)| \leq C_4(m, \rho)|z|^{m\rho/2}e^{\operatorname{Re} x \operatorname{Re} z - (\frac{k\pi}{2} - |\operatorname{Im} x|)|\operatorname{Im} z|} \tag{72}$$

Let

$$\delta = \min(-\operatorname{Re} x, \frac{k\pi}{2} - |\operatorname{Im} x|). \tag{73}$$

The inequality (72) implies

$$|h(x, z)| \le C_4(m, \rho)|z|^{m\rho/2} e^{-\delta|z|} \tag{74}$$

and hence

$$|h(x, z)| \le C_4(m, \rho)M''|z|^{-2} \tag{75}$$

where

$$M'' = (\frac{2 + m\rho/2}{\delta})^{2+m\rho/2}. \tag{76}$$

The argument then proceeds as before with M'' in place of M'.

Finally we prove 3. Let

$$\phi = \frac{1}{2l}(\frac{k\pi}{2} - X_i). \tag{77}$$

If $|\arg z| > \frac{\pi}{2} - \phi$ then

$$e^{(l|\arg z|+(k-l)\frac{\pi}{2}-X_i)|\operatorname{Im} z|} \le e^{-\phi|\operatorname{Im} z|}. \tag{78}$$

Let ξ be any real number greater than λ such that

$$|\lambda - a_j - n| \ge \frac{1}{4m} \tag{79}$$

for all j, n, and let

$$H_1 = \eta \cot \phi. \tag{80}$$

If $|\operatorname{Im} z| > H_1$ and $0 \le \operatorname{Re} z \le \eta$ then $|\arg z| > \frac{\pi}{2} - \phi$. In this region then (10) implies

$$|h(x, z)| \le C_4(m, \rho)|z|^{m\rho/2+l\eta} e^{-\phi|\operatorname{Im} z|} \le C_4(m, \rho)|z|^{m\rho/2+l\eta} e^{-\phi \sec \phi|z|}. \tag{81}$$

By (23) of the appendix

$$|z|^{m\rho/2+l\xi} e^{-\phi \sec \phi|z|} \le M''''|z|^{-2} \tag{82}$$

where

$$M'''' = \left(\frac{2 + m\rho/2 + l\eta}{\phi \sec \phi} \right)^{2+m\rho/2} \tag{83}$$

Let C_- be a contour which follows the imaginary axis from $-i\infty$ to $-i\eta$, then continues to be a line segment to $\xi - i\eta$, and finally returns to $-i\infty$ along the line $\operatorname{Re} z = \xi$. If $\eta > H_1$ then (82) implies

$$\left| \int_{C_-} h(x, z)\, dz \right| \le 3\pi C_4(m, \rho) M'''' \eta^{-1}, \tag{84}$$

and the right hand side tends to zero as η tends to infinity. It follows from condition 4 of Theorem 9, though, that the integral is independent of η. Thus

$$\int_{C_-} h(x, z)\, dz = 0. \tag{85}$$

Let C_+ be a contour which follows the imaginary axis from $i\infty$ to $i\eta$, then continues to be a line segment to $\xi + i\eta$, and finally returns to $+i\infty$ along the line $\operatorname{Re} z = \xi$. An entirely analogous argument shows that

$$\int_{C_+} h(x, z)\, dz = 0. \tag{86}$$

We conclude that

$$\int_{C_{\xi,\eta}} h(x, z)\, dz = \int_{\xi-i\infty}^{\xi+i\infty} h(x, z)\, dz, \tag{87}$$

where the integral on the right is the integral along the line $\operatorname{Re} z = \xi$. On this line

$$|h(x, z)| \le C_4(m, \rho)|z|^{l\xi+m\rho/2} e^{\operatorname{Re} x \operatorname{Re} z - l\xi - (l|\arg z| + (k-l)\frac{\pi}{2} - X_i)|\operatorname{Im} z|}. \tag{88}$$

and

$$\left| \int_{\xi-i\infty}^{\xi+i\infty} h(x, z)\, dz \right| \le K e^{\xi \operatorname{Re} x} \le K e^{\lambda \operatorname{Re} x} \tag{89}$$

where

$$K = C_4(m, \rho) \int_{\xi-i\infty}^{\xi+i\infty} |z|^{l\xi+m\rho/2} e^{-l\xi - (l|\arg z| + (k-l)\frac{\pi}{2} - X_i)|\operatorname{Im} z|} |dz|. \tag{90}$$

This establishes 3 and completes the proof of Theorem 27.

Now that we have Theorem 27 available it is a fairly simple matter to prove theorems 25 and 26. We begin with 25.

Proof. Define

$$f(t) = -\frac{1}{2\pi i} \int_C \frac{\Gamma(z+a)\Gamma(z+b)\Gamma(c)}{\Gamma(a)\Gamma(b)\Gamma(z+c)} \Gamma(-z) e^{z\log(-t)} \, dz. \tag{91}$$

The contour is chosen so that the poles of $\Gamma(-z)$ lie to the right while those of $\Gamma(z+a)$ and $\Gamma(z+b)$ lie to the left. The integral is of the form (1) with $k = 2$, $l = 0$. Part 1 of Theorem 27 shows that the integral converges. We use part 2 to determine the expansion about $t = 0$. Let C''' be a contour which follows $C_{N+\alpha,N+\alpha}$ from $-i(N+\alpha)$ to $+i(N+\alpha)$ and then returns along C. Thus

$$\int_C = \int_{C'''} + \int_{C_{N+\alpha,N+\alpha}}. \tag{92}$$

Theorem 22 gives the representation

$$\frac{1}{2\pi i} \int_{C'''} \frac{\Gamma(z+a)\Gamma(z+b)\Gamma(c)}{\Gamma(a)\Gamma(b)\Gamma(z+c)} \Gamma(-z) e^{z\log(-t)} \, dz$$

$$= \sum \operatorname{Ind}(C''', w) \operatorname{Res}_{z=w} \frac{\Gamma(z+a)\Gamma(z+b)\Gamma(c)}{\Gamma(a)\Gamma(b)\Gamma(z+c)} \Gamma(-z) e^{z\log(-t)}. \tag{93}$$

By part 2 of Theorem 27

$$\lim_{N\to\infty} \int_{C_{N+\alpha,N+\alpha}} \frac{\Gamma(z+a)\Gamma(z+b)\Gamma(c)}{\Gamma(a)\Gamma(b)\Gamma(z+c)} \Gamma(-z) e^{z\log(-t)} \, dz = 0, \tag{94}$$

for $|t| < 1$, and hence

$$f(t) = \lim_{N\to\infty} \sum \operatorname{Ind}(C''', w) \operatorname{Res}_{z=w} \frac{\Gamma(z+a)\Gamma(z+b)\Gamma(c)}{\Gamma(a)\Gamma(b)\Gamma(z+c)} \Gamma(-z) e^{z\log(-t)}. \tag{95}$$

As N tends to infinity all the poles of $\Gamma(-z)$ are included in the sum, each with index 1. We conclude that

$$f(t) = \sum_{n=0}^{\infty} \operatorname{Res}_{z=n} \frac{\Gamma(z+a)\Gamma(z+b)\Gamma(c)}{\Gamma(a)\Gamma(b)\Gamma(z+c)} \Gamma(-z) e^{z\log(-t)}. \tag{96}$$

The residues were evaluated in Section 2.8. We find

$$f(t) = F(a, b; c; t). \tag{97}$$

To get the expansion about infinity we make the change of variable $z \to -z$ in the integral defining $f(t)$. More precisely, let $\tilde{C}$ be the contour corresponding to $\tilde{\gamma}$, defined by

$$\tilde{\gamma}(t) = -\gamma(-t). \tag{98}$$

Then

$$f(t) = \frac{1}{2\pi i} \int_{\tilde{C}} \frac{\Gamma(-z+a)\Gamma(-z+b)\Gamma(c)}{\Gamma(a)\Gamma(b)\Gamma(-z+c)} \Gamma(z) e^{z \log(-t^{-1})}, \tag{99}$$

an integral of the same type. Using part 2 of Theorem 27 again we find

$$f(t) = -\sum \frac{\Gamma(z+a)\Gamma(z+b)\Gamma(c)}{\Gamma(a)\Gamma(b)\Gamma(z+c)} \Gamma(-z) e^{z \log(-t)} \tag{100}$$

for $|t| > 1$, where the sum is now over the poles to the left of C. The residues were evaluated in Section 2.8. Inserting them into the preceding equation gives exactly equation (23) of Section 2.8, so the proof of Theorem 25 is complete.

Next we prove Theorem 26.

Proof. We define

$$f(s) = -\frac{1}{2\pi i} \int_C \frac{\Gamma(z+p)\Gamma(z+p+q)}{\Gamma(p)\Gamma(p+q)} \Gamma(-z) e^{z \log s^{-1}} \, dz \tag{101}$$

The contour is chosen so that the poles of $\Gamma(-z)$ lie to the right while those of $\Gamma(z+p)$ and $\Gamma(z+p+q)$ lie to the left. The integral is of the form (1) with $k = 3, l = 1$. Part 1 of Theorem 27 shows that the integral converges.

Part 3 gives the expansion about $s = \infty$. It shows that

$$f(s) = \sum_{n<\xi} \operatorname*{Res}_{z=n} \frac{\Gamma(z+p)\Gamma(z+p+q)}{\Gamma(p)\Gamma(p+q)} \Gamma(-z) e^{z \log s^{-1}} + O(|s|^{-N-1}), \tag{102}$$

For some $\xi < N + 1$. We evaluated the residues in Section 2.8. They were

$$(-1)^n \frac{(p)_n (p+q)_n}{(1)_n} s^{-n}. \tag{103}$$

The terms with $n > N$ may be absorbed into the error term, leaving

$$f(s) = \sum_{n \le N} (-1)^n \frac{(p)_n (p+q)_n}{(1)_n} s^{-n} + O(|s|^{-N-1}), \tag{104}$$

which shows that $f(s)$ has the same asymptotic expansion at infinity as $J(p, q, s)$.

Part 2 gives the expansion about $s = 0$. Defining

$$\tilde{\gamma}(t) = -\gamma(-t). \tag{105}$$

we see that

$$f(s) = \frac{1}{2\pi i} \int_{\tilde{C}} \frac{\Gamma(-z+p)\Gamma(-z+p+q)}{\Gamma(p)\Gamma(p+q)} \Gamma(z) e^{z \log s} \, dz, \tag{106}$$

an integral of the form (1), with $k = 3$, $l = -1$. An argument almost identical to that given for the hypergeometric function gives the expansion

$$\frac{\Gamma(q)}{\Gamma(p+q)} \sum_{n=0}^{\infty} \frac{(p)_n}{(1)_n(1-q)_n} s^n + \frac{\Gamma(-q)}{\Gamma(p)} \sum_{n=0}^{\infty} \frac{(p+q)_n}{(1)_n(1+q)_n} s^{q+n}, \tag{107}$$

which agrees with that given in Theorem 26.

Exercises

1. Prove that

$$e^{-u} = -\frac{1}{2\pi 1} \int_C \Gamma(z) u^z \, dz$$

for $-\frac{\pi}{2} < \arg u < \frac{\pi}{2}$, where C is a contour which foolws the imaginary axis form $-i\infty$ to $+i\infty$ except for a brief detour pass to the left of the poles of $\Gamma(z)$.

2. Prove that $J(p, q, s)$ has the integral representation

$$\frac{1}{2\pi i} \int_{\tilde{C}} \frac{\Gamma(-z+p)\Gamma(-z+p+q)}{\Gamma(p)\Gamma(p+q)} \Gamma(z) e^{z \log s} \, dz.$$

Formally this follows from inserting the integral given in the preceding exercise for the exponential function into the definite integral defining $J(p, q, s)$ and then reversing the order of integration. Show that reversing the order of integration is permissible. Note that you need only prove this when all parameters are positive and real, since the equality can then be extended by the rigidity principle.

3. Extend the argument which gave the inequality (17) to show that Stirling's series gives an asymptotic expansion for $\Gamma(\zeta)$ in the region given by (15) and (16).

4. Derive the asymptotic expansion of $\Gamma(z - a)$ in powers of z^{-1}.

5. What is the expansion of

$$\sum_{n=0}^{\infty} \frac{(a)_n}{(1)_n(b)_n} z^n$$

at infinity and in what region is it valid?

Chapter 3

Elliptic and Modular Functions

In the chapter we will examine theta functions, elliptic, and modular functions. We will study theta functions in great detail. Since the elliptic and modular functions are simply rational expressions in theta functions, they can be disposed of quite quickly.

3.1 Theta Functions

In this section we will meet the theta functions and Eisenstein series, both of which play an important role in the analytic theory of numbers. The most important property of these functions is their dependence on the variable τ, which we treat in the next section. Here we concentrate on the dependence on the variable ζ.

The classical Jacobi[1] theta functions are defined by the series expansions

$$\theta_0(\tau, \zeta) = \sum_{k=-\infty}^{\infty} (-1)^k q^{k^2} z^k$$

$$\theta_1(\tau, \zeta) = i \sum_{k=-\infty}^{\infty} (-1)^k q^{(k-\frac{1}{2})^2} z^{k-\frac{1}{2}}$$

$$\theta_2(\tau, \zeta) = \sum_{k=-\infty}^{\infty} q^{(k-\frac{1}{2})^2} z^{k-\frac{1}{2}}$$

---

[1]In fact the first appearance of θ_3 in print is due to Fourier, who used it in his study [29] of heat conduction, the same source from which we have Fourier series. Fourier considered only real ζ and τ of the form it with t real. Equation (4) below then becomes, for appropriate choice of units, the heat equation. θ_3 is the fundamental solution with periodic boundary conditions. All four functions are older though. They appear in Gauss' notes, but he published nothing and does not seem to have told anyone of his work until after Jacobi's investigations. Closely related functions appeared already in Euler's work on partitions on integers. See, for example, the section "Partitio Numerorum" of [26].

J. Stalker, *Complex Analysis: Fundamentals of the Classical Theory of Functions*,
Modern Birkhäuser Classics, DOI 10.1007/978-0-8176-4919-7_3,
© Birkhäuser Boston, a part of Springer Science + Business Media, LLC 2009

$$\theta_3(\tau, \zeta) \;=\; \sum_{k=-\infty}^{\infty} q^{k^2} z^k \tag{1}$$

where $q = e^{\pi i \tau}$ and $z = e^{2\pi i \zeta}$. Fractional powers are interpreted in the obvious way,

$$q^\alpha = e^{\pi i \alpha \tau}, \quad z^\beta = e^{2\pi i \beta \zeta}. \tag{2}$$

These series converge locally uniformly for $\operatorname{Im}\tau > 0$, $\zeta \in \mathbf{C}$ and thus represent analytic functions of both arguments there.

There are almost as many rival notations for theta functions as there are books on the subject. Here we follow Kronecker's notation except where specifically indicated. This in turn is based on Jacobi's later notation. We write θ_0 in place of Kronecker's θ.

THEOREM 28 (ELEMENTARY PROPERTIES OF THETA FUNCTIONS)
Theta functions satisfy the following translation relations

$$\theta_0(\tau, \zeta + \tfrac{1}{2}) = \theta_3(\tau, \zeta) \qquad \theta_0(\tau, \zeta + \tfrac{\tau}{2}) = iq^{-1/4}z^{-1/2}\theta_1(\tau, \zeta) \tag{3}$$

$$\theta_1(\tau, \zeta + \tfrac{1}{2}) = \theta_2(\tau, \zeta) \qquad \theta_1(\tau, \zeta + \tfrac{\tau}{2}) = iq^{-1/4}z^{-1/2}\theta_0(\tau, \zeta)$$

$$\theta_2(\tau, \zeta + \tfrac{1}{2}) = -\theta_1(\tau, \zeta) \qquad \theta_2(\tau, \zeta + \tfrac{\tau}{2}) = q^{-1/4}z^{-1/2}\theta_3(\tau, \zeta)$$

$$\theta_3(\tau, \zeta + \tfrac{1}{2}) = \theta_0(\tau, \zeta) \qquad \theta_3(\tau, \zeta + \tfrac{\tau}{2}) = q^{-1/4}z^{-1/2}\theta_2(\tau, \zeta),$$

the partial differential equation

$$4\pi i \frac{\partial}{\partial \tau}\theta_j(\tau, \zeta) = \frac{\partial^2}{\partial \zeta^2}\theta_j(\tau, \zeta), \tag{4}$$

and the parity relations

$$\theta_0(\tau, -\zeta) = \theta_0(\tau, \zeta) \tag{5}$$
$$\theta_1(\tau, -\zeta) = -\theta_1(\tau, \zeta)$$
$$\theta_2(\tau, -\zeta) = \theta_2(\tau, \zeta)$$
$$\theta_3(\tau, -\zeta) = \theta_3(\tau, \zeta).$$

Proof.   All of these are proved by rearranging power series, the formal operations being justified by absolute convergence.

There is no standard notation for the logarithmic derivatives of theta functions. We define

$$f_j(\tau, \zeta) = \frac{\frac{\partial}{\partial \zeta}\theta_j(\tau, \zeta)}{\theta_j(\tau, \zeta)}. \tag{6}$$

THEOREM 29 (LOCATION OF ZEROES)
$\theta_j(\tau, \zeta) = 0$ *if and only if* $\zeta = m\tau + n$ *where the conditions on m and n are as follows:*

- *For* $j = 0$, $m - \frac{1}{2} \in \mathbf{Z}$, $n \in \mathbf{Z}$.

- *For* $j = 1$, $m \in \mathbf{Z}$, $n \in \mathbf{Z}$.

- *For* $j = 2$, $m \in \mathbf{Z}$, $n - \frac{1}{2} \in \mathbf{Z}$.

- *For* $j = 3$, $m - \frac{1}{2} \in \mathbf{Z}$, $n - \frac{1}{2} \in \mathbf{Z}$.

These zeroes are all of order 1.

Proof.   We will prove this theorem only when $j = 1$. The other cases follow at once from the translation relations (3). First we must check that these points are zeroes.

$$\theta_1(\tau, 0) = 0 \tag{7}$$

follows from the parity relation (5). The nontrivial part of the theorem is that there are no other zeros.

Next we check that there is no τ such that $\theta_1(\tau, \zeta)$ vanishes for all ζ. If there were such a τ then

$$\frac{\partial^{2l}}{\partial \zeta^{2l}}\theta_1(\tau, \zeta) = 0$$

and hence, by (4),

$$\frac{\partial^l}{\partial \tau^l}\theta_1(\tau, \zeta) = 0$$

for all l, ζ and this τ. Theorem 18 would then show that

$$\theta_1(\tau, \zeta) = 0$$

for all τ with Im $\tau > 0$. It is clear though from (1) that

$$\lim_{\tau \to i\infty} e^{\pi i \tau/4}\theta_1(\tau, \zeta) = 2 \sin \pi \zeta, \tag{8}$$

so there can be no such τ. Another application of the rigidity theorem shows that, for any fixed τ, the zeroes of θ_1 are discrete.

The next step is to use the zero counting theorem from Section 2.6 to count the zeroes in a parallelogram. Since these zeroes are already known to be discrete, we can therefore find a ζ_0 such that the counterclockwise contour C with straight edges and corners at $\zeta_0 \pm \frac{\tau}{2} \pm \frac{1}{2}$ avoids the zeroes of θ_1. Theorem 23 then shows that the number of zeroes inside C, counting multiplicities, is given by the contour integral

$$\frac{1}{2\pi i} \int_C f_1(\tau, \xi)\, d\xi.$$

The translation relations

$$f_1(\tau, \zeta + 1) = f_1(\tau, \zeta) \quad f_1(\tau, \zeta + \tau) = f_1(\tau, \zeta) - 2\pi i, \tag{9}$$

which follow from (3), allow us to calculate

$$\frac{1}{2\pi i} \int_C f_1(\tau, \xi)\, d\xi = 1. \tag{10}$$

Thus there is only one zero inside C, the one we have already found, and the proof of the theorem is complete.

Exercises

1. Prove that if f is continuous on $\mathbf{R}$ and periodic of period L,

$$f(x + L) = f(x),$$

   then

$$u(x, t) = \frac{1}{L} \int_0^L \theta_3\left(\frac{4\pi i L^2}{k}, L\xi\right) f(x - \xi)\, d\xi.$$

   is a solution of the heat equation

$$\frac{\partial u}{\partial t} = k \frac{\partial^2 u}{\partial x^2}$$

   satisfying the initial condition

$$\lim_{t \to 0^+} u(x, t) = f(x).$$

   This result first appears in Fourier's [29].

2. Prove that

$$\int_0^\infty \theta_3(iy, 0) y^{s-1}\, dy = \Gamma(s)(2\pi)^s \zeta(2s).$$

3. Prove that

$$\theta_2(4\tau, \zeta) = \frac{1}{2}\theta_3(\tau, \frac{1}{2}\zeta) - \frac{1}{2}\theta_0(\tau, \frac{1}{2}\zeta)$$

and

$$\theta_3(4\tau, \zeta) = \frac{1}{2}\theta_3(\tau, \frac{1}{2}\zeta) + \frac{1}{2}\theta_0(\tau, \frac{1}{2}\zeta).$$

3.2 Eisenstein Series

We can use Theorem 29 to compute the Laurent series of f_1 by the same method we used to compute the Laurent series of the cotangent in Section 2.7. We write the Laurent expansion about $\zeta = 0$ in the form

$$f_1(\tau, \zeta) = -\sum_{l=0}^{\infty} E_l(\tau)\zeta^{l-1} \tag{1}$$

and call the coefficient $E_l(\tau)$ the *Eisenstein series*[2] of weight l. The parity relation (5) shows that

$$f_1(\tau, -\zeta) = -f_1(\tau, \zeta), \tag{2}$$

so the coefficients of even powers vanish,

$$f_1(\tau, \zeta) = -\sum_{k=0}^{\infty} E_{2k}(\tau)\zeta^{2k-1}. \tag{3}$$

Let $C_{M,N}$ be a counterclockwise-oriented contour with straight sides and corners at $\pm(M + \frac{1}{2})\tau \pm (N + \frac{1}{2})$. An easy residue calculation gives

$$\int_{C_{M,N}} f_1(\tau, \xi)\xi^{-2k} d\xi = \sum_{\substack{-M \le m \le M \\ -N \le n \le N \\ (m, n) \ne (0, 0)}} (m\tau + n)^{-2k} - E_{2k}(\tau). \tag{4}$$

For the remainder of this section any sum over m and n which is otherwise unspecified will be over the same range as above. f_1 is bounded on $C_{0,0}$ by Theorem L of the Appendix, so the translation relations imply

$$|f_1(\tau, \xi)| = O(M) \tag{5}$$

---

[2]Eisenstein uses the notation $(l, *)$, which is unsuitable for our purposes because the dependence on τ is not explicit. These series have no fixed notation in Jacobi and Kronecker, but the symbol E_l has since become standard. Modern usage would deny to E_0 and E_2 the name Eisenstein series, but Eisenstein himself made no such distinction.

on $C_{M,N}$, while

$$|\xi|^{-2k} = O(\min(M, N)) \tag{6}$$

and

$$\text{length } C_{M,N} = O(M + N). \tag{7}$$

None of these estimates is uniform in τ, but they will not need to be. Using the estimate (12) from Section 2.1, we obtain the estimate

$$\left| \int_{C_{M,M}} f_1(\tau, \xi) \xi^{-2k} \, d\xi \right| = O(M^{2-2k}). \tag{8}$$

If $k > 1$ we can let M tend to infinity and we find

$$E_{2k}(\tau) = \lim_{M \to \infty} \sum_{\substack{-M \le m, n \le M \\ (m, n) \ne (0, 0)}} (m\tau + n)^{-2k} \tag{9}$$

or, since the sum converges absolutely,

$$E_{2k}(\tau) = \sum_{\substack{-\infty \le m, n \le \infty \\ (m, n) \ne (0, 0)}} (m\tau + n)^{-2k}. \tag{10}$$

We emphasize that for $k = 1$ equation (10) is meaningless because the sum is not absolutely convergent. Equation (9) is meaningful but false for $k = 1$. Computing E_0 is trivial. E_0 is minus the residue of $f_1(\zeta)$ at the pole $\zeta = 0$, which we calculated in (10), so

$$E_0(\tau) = -1. \tag{11}$$

E_2, however, will take a great deal of effort. We must be very careful with the order of summation, for reasons which will become clear at the end of the section.

We now begin computing E_2 by the calculus of residues. To simplify the formulae which follow we set

$$\mu = (M + \frac{1}{2})\tau. \tag{12}$$

Clearly

$$\int_{C_{M,N}} - \int_{C_{M,N'}} = \int_{C_{\text{left}}} + \int_{C_{\text{right}}} \tag{13}$$

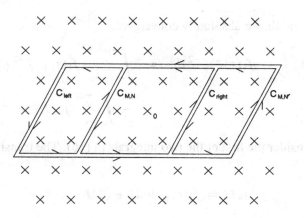

Figure 3.1: Contours for evaluation of Eisenstein series.

where C_{left} and C_{right} are polygonal contours, oriented either clockwise or counterclockwise according to whether $N < N'$ or $N > N'$. The former has corners at $\pm\mu - (N + \frac{1}{2})$ and $\pm\mu - (N' + \frac{1}{2})$, while the latter has corners at $\pm\mu + (N + \frac{1}{2})$ and $\pm\mu + (N' + \frac{1}{2})$. The case $\tau = 1 + \frac{1}{2}$, $M = 1$, $N = 1$, $N' = 3$ is illustrated in the picture. Letting $C = C_{\text{left}}$ or $C = C_{\text{right}}$ in (52) of Section 2.9. we find

$$\int_C f_1(\tau, \xi)\xi^{-2}\, d\xi = O(1/\min(N, N')). \tag{14}$$

The estimate is not uniform in M or τ, but it doesn't need to be. From (13) and (14) we see that the sequence

$$\int_{C_{M,N}} f_1(\tau, \xi)\xi^{-2}\, d\xi$$

is Cauchy, and hence convergent, as N tends to infinity with M fixed. By (51) of Section 2.9 the contributions from the right and left sides of the polygon tend to zero, so

$$\lim_{N\to\infty} \int_{C_{M,N}} f_1(\tau, \xi)\xi^{-2}\, d\xi = \lim_{N\to\infty} \left(\int_{-\mu-N-\frac{1}{2}}^{-\mu+N+\frac{1}{2}} f_1(\tau, \xi)\xi^{-2}\, d\xi \right. \tag{15}$$

$$\left. - \int_{\mu-N-\frac{1}{2}}^{\mu+N+\frac{1}{2}} f_1(\tau, \xi)\xi^{-2}\, d\xi \right)$$

or, since the integrals are absolutely convergent,

$$\lim_{N \to \infty} \int_{C_{M,N}} f_1(\tau, \xi) \xi^{-2} \, d\xi = \int_{-\mu-\infty}^{-\mu+\infty} f_1(\tau, \xi) \xi^{-2} \, d\xi \tag{16}$$

$$- \int_{\mu-\infty}^{\mu+\infty} f_1(\tau, \xi) \xi^{-2} \, d\xi.$$

We now consider the first of the two integrals in (16). The translation relations imply

$$f_1(\tau, \xi) = f_0(\tau, \xi + \mu) + (2M + 1)\pi i \tag{17}$$

so

$$\int_{-\mu-\infty}^{-\mu+\infty} f_1(\tau, \xi) \xi^{-2} \, d\xi = \int_{-\mu-\infty}^{-\mu+\infty} f_0(\tau, \xi + \mu) \xi^{-2} \, d\xi \tag{18}$$

$$+ (2M + 1)\pi i \int_{-\mu-\infty}^{-\mu+\infty} \xi^{-2} \, d\xi$$

The second of these integrals is zero. We estimate the first using (51) of Section 2.9. The line segment is infinite, but this doesn't affect the argument. We find

$$\left| \int_{-\mu-\infty}^{-\mu+\infty} f_1(\tau, \xi) \xi^{-2} \, d\xi \right| \le \pi \frac{\sup_{\xi \in (-\mu-\infty, -\mu+\infty)} |f_0(\tau, \xi + \mu)|}{\operatorname{Im} \mu} \tag{19}$$

or, setting $\eta = \xi + \mu$,

$$\left| \int_{-\mu-\infty}^{-\mu+\infty} f_1(\tau, \xi) \xi^{-2} \, d\xi \right| \le \pi \frac{\sup_{\eta \in (-\infty, +\infty)} |f_0(\tau, \eta)|}{\operatorname{Im} \mu}. \tag{20}$$

The numerator is finite and independent of M while the denominator is

$$(M + \frac{1}{2}) \operatorname{Im} \tau,$$

so

$$\int_{-\mu-\infty}^{-\mu+\infty} f_1(\tau, \xi) \xi^{-2} \, d\xi = O(M^{-1}). \tag{21}$$

Similarly

$$\int_{\mu-\infty}^{\mu+\infty} f_1(\tau, \xi) \xi^{-2} \, d\xi = O(M^{-1}) \tag{22}$$

and hence

$$\lim_{M\to\infty} \lim_{N\to\infty} \int_{C_{M,N}} f_1(\tau, \xi)\xi^{-2}\,d\xi = 0. \tag{23}$$

The order in which the limits above are taken—M tends to infinity only after N tends to infinity—is important. Most, but not all, of the limits which follow will be taken in the same order. We therefore adopt the notational convention that any otherwise unspecified limit is to be evaluated in this order,

$$\lim = \lim_{M\to\infty} \lim_{N\to\infty} .$$

Equation (4) shows that

$$E_2(\tau) = \lim \sum (m\tau + n)^{-2} \tag{24}$$

where we are employing our convention that all otherwise unspecified sums have the same range as (4). If we want formulae which hold for all $k \geq 1$ then we can write

$$E_{2k}(\tau) = \lim \sum (m\tau + n)^{-2k} \tag{25}$$

since the sum converges absolutely when $k > 1$.

We can use the results of Section 2.7 to express E_{2k} in terms of q. Writing

$$E_{2k}(\tau) = \lim_{N\to\infty} \sum_{\substack{-N \leq n \leq N \\ n \neq 0}} n^{-2k} \tag{26}$$

$$+ \lim \sum_{\substack{-M \leq m \leq M \\ -N \leq n \leq N \\ m \neq 0}} (m\tau + n)^{-2k}$$

we use equation (26) to evaluate the first term and Exercise 5 to evaluate the second. This gives

$$E_{2k}(\tau) = (-1)^{k+1}\frac{(2\pi)^{2k}}{(2k-1)!}[\frac{B_{2k}}{2k} - 2\sum_{m=1}^{\infty}\sum_{d=1}^{\infty} d^{2k-1}q^{2dm}]. \tag{27}$$

The double sum converges absolutely, so we may rearrange terms so as to group those with $dm = n$. This produces

$$E_{2k}(\tau) = (-1)^{k+1}\frac{(2\pi)^{2k}}{(2k-1)!}[\frac{B_{2k}}{2k} - 2\sum_{n=1}^{\infty} \sigma_{2k-1}(n)q^{2n}], \tag{28}$$

where

$$\sigma_{2k-1}(n) = \sum_{d|n} d^{2k-1}. \tag{29}$$

The sum runs over positive divisors of n. Just as trigonometric identities implied nontrivial identities among sums in Section 2.7 we can use theta function identities to derive information about these sums of powers of divisors. First, of course, we will need to find some nontrivial theta function identities.

The Eisenstein series satisfy a remarkable functional equation. If we set[3]

$$G_{2k}(\omega_1, \omega_2) = \lim \sum (m\omega_2 + n\omega_1)^{-2k} \tag{30}$$

then

$$G_{2k}(\omega_1, \omega_2) = \omega_1^{-2k} E_{2k}\left(\frac{\omega_2}{\omega_1}\right). \tag{31}$$

Suppose ω_1' and ω_2' are related to ω_1 and ω_2 by

$$\begin{pmatrix} \omega_2' \\ \omega_1' \end{pmatrix} = \begin{pmatrix} a & b \\ c & d \end{pmatrix} \begin{pmatrix} \omega_2 \\ \omega_1 \end{pmatrix} \tag{32}$$

where $a, b, c, d \in \mathbf{Z}$ and $ad - bc = 1$. Then the set Λ consisting of all of ω of the form $m\omega_2 + n\omega_1$ coincides with the set of ω of the form $m'\omega_2' + n'\omega_1'$ since

$$\begin{pmatrix} m & n \end{pmatrix} \begin{pmatrix} \omega_2 \\ \omega_1 \end{pmatrix} = \begin{pmatrix} m' & n' \end{pmatrix} \begin{pmatrix} \omega_2' \\ \omega_1' \end{pmatrix} \tag{33}$$

where

$$\begin{pmatrix} m' & n' \end{pmatrix} = \begin{pmatrix} m & n \end{pmatrix} \begin{pmatrix} d & -b \\ -c & a. \end{pmatrix}. \tag{34}$$

Equation (31) remains valid if ω_1 and ω_2 are replaced by ω_1' and ω_2', so

$$G_{2k}(\omega_1', \omega_2') = \omega_1'^{-2k} E_{2k}\left(\frac{\omega_2'}{\omega_1'}\right). \tag{35}$$

Both $G_2(\omega_1, \omega_2)$ and $G_2(\omega_1', \omega_2')$ are sums of ω^{-2k} over the set $\Lambda - \{0\}$. They differ only in the order of summation. For $k > 1$ the sum is absolutely convergent, so the sum is independent of the order,

$$G_{2k}(\omega_1', \omega_2') = G_{2k}(\omega_1, \omega_2) \tag{36}$$

---

[3]The notation G_{2k} is widespread, but not universal.

or, in terms of the Eisenstein series,

$$\left(\frac{\omega_1'}{\omega_1}\right)^{-2k} E_{2k}\left(\frac{\omega_2'}{\omega_1'}\right) = E_{2k}(\frac{\omega_2}{\omega_1}). \tag{37}$$

Choosing ω_1 and ω_2 so that $\frac{\omega_2}{\omega_1} = \tau$ we find

$$(c\tau + d)^{-2k} E_{2k}\left(\frac{a\tau + b}{c\tau + d}\right) = E_{2k}(\tau). \tag{38}$$

Functions satisfying a functional equation of this kind are called modular forms. We will meet more of them later. None of equations (36), (37), or (38) holds for $k = 1$.

To find the correct functional equation for $k = 1$, we will need the auxiliary function[4]

$$\phi(x, \omega_1, \omega_2) = \omega_1^{-1} f_1(\frac{\omega_2}{\omega_1}, \frac{x}{\omega_1}). \tag{39}$$

The translation relations (9) of the preceding section imply the relations

$$\phi(x + \omega_1, \omega_1, \omega_2) = \phi(x, \omega_1, \omega_2) \tag{40}$$

and

$$\phi(x + \omega_2, \omega_1, \omega_2) = \phi(x, \omega_1, \omega_2) - \frac{2\pi i}{\omega_1} \tag{41}$$

for ϕ. Using (3) and (31) we find the Laurent expansion

$$\phi(x, \omega_1, \omega_2) = - \sum_{k=0}^{\infty} G_{2k}(\omega_1, \omega_2) x^{2k-1}, \tag{42}$$

from which

$$\phi(x, \omega_1', \omega_2') - \phi(x, \omega_1, \omega_2) = -[G_2(\omega_1', \omega_2') - G_2(\omega_1, \omega_2)]x. \tag{43}$$

We can use this equation to compute $G_2(\omega_1', \omega_2') - G_2(\omega_1, \omega_2)$. For any $x, y \in \mathbf{C}$ we have

$$G_2(\omega_1', \omega_2') - G_2(\omega_1, \omega_2) = \frac{\phi(y, \omega_1, \omega_2) - \phi(x, \omega_1, \omega_2)}{y - x}$$

---

[4]This function is called $(1, x)$ by Eisenstein, a notation which we cannot follow, because it does not make the dependence on ω_1 and ω_2 explicit. Weil calls the same function $E_1(x; \omega_1, \omega_2)$ but we are already using the symbol E_1 as a synonym for 0.

$$-\frac{\phi(y, \omega_1', \omega_2') - \phi(x, \omega_1', \omega_2')}{y - x}. \qquad (44)$$

We apply this with

$$y = x + \omega_1' = x + c\omega_2 + d\omega_1, \qquad (45)$$

obtaining

$$G_2(\omega_1', \omega_2') - G_2(\omega_1, \omega_2) = -\frac{2\pi i c}{\omega_1 \omega_1'}. \qquad (46)$$

In terms of the Eisenstein series this gives the functional equation

$$(c\tau + d)^{-2} E_2(\frac{a\tau + b}{c\tau + d}) = E_2(\tau) - \frac{2\pi i c}{c\tau + d}. \qquad (47)$$

We can now use equation (46) to put (43) in the more attractive form

$$\phi(x, \omega_1', \omega_2') = \phi(x, \omega_1, \omega_2) + \frac{2\pi i c x}{\omega_1 \omega_1'}. \qquad (48)$$

Choosing x, ω_1, and ω_2 so that $\frac{\omega_2}{\omega_1} = \tau$ and $\frac{x}{\omega_1} = \zeta$ we find

$$(c\tau + d)^{-1} f_1(\frac{a\tau + b}{c\tau + d}, \frac{\zeta}{c\tau + d}) = f_1(\tau, \zeta) + \frac{2\pi i c\zeta}{c\tau + d} \qquad (49)$$

Since f_1 is just the logarithmic derivative of θ_1 we can integrate to find the functional equation

$$\frac{\theta_1(\frac{a\tau + b}{c\tau + d}, \frac{\zeta}{c\tau + d})}{\theta_1(\tau, \zeta)} = \lim_{\zeta \to 0} \frac{\theta_1(\frac{a\tau + b}{c\tau + d}, \frac{\zeta}{c\tau + d})}{\theta_1(\tau, \zeta)} e^{\pi i \frac{c\zeta^2}{c\tau + d}} \qquad (50)$$

for θ_1. Using L'Hôpital's rule to evaluate the limit, we find

$$\frac{\theta_1(\frac{a\tau + b}{c\tau + d}, \frac{\zeta}{c\tau + d})}{\theta_1(\tau, \zeta)} = (c\tau + d)^{-1} \frac{\theta_1'(\frac{a\tau + b}{c\tau + d}, 0)}{\theta_1'(\tau, 0)} e^{\pi i \frac{c\zeta^2}{c\tau + d}}, \qquad (51)$$

where primes denote derivatives with respect to the second argument. This is a crude form of the transformation formulae for theta functions. We will see a more precise form in the next section.

We can now see why the argument that gave equation (23) was so delicate. The order in which the limits are evaluated is important. Just for the next calculation let us denote by $\lim'$ the limit taken in the opposite order,

$$\lim' = \lim_{N \to \infty} \lim_{M \to \infty},$$

in words, the limit as M tends to infinity and then N tends to infinity. We compute

$$
\begin{aligned}
\lim{}' \frac{1}{2\pi i} \int_{C_{M,N}} f_1(\tau, \xi) \xi^{-2}\, d\xi &= \lim{}' \sum (m\tau + n)^{-2} - E_2(\tau) \qquad (52) \\
&= \lim{}' \sum (m\tau - n)^{-2} - E_2(\tau) \\
&= \lim \sum (n\tau - m)^{-2} - E_2(\tau) \\
&= \tau^{-2} \lim \sum (m\frac{-1}{\tau} + n)^{-2} - E_2(\tau) \\
&= \tau^{-2} E_2(\frac{-1}{\tau}) - E_2(\tau) \\
&= \frac{2\pi i}{\tau} \\
&\neq 0.
\end{aligned}
$$

We could not therefore have shifted the contours in the opposite order.

We now derive product representations for the theta functions. We begin by finding a sum representation for ϕ. For $|x|$ sufficiently small,

$$
\begin{aligned}
\phi(x, \omega_1, \omega_2) &= x^{-1} - G_2(\omega_1, \omega_2)x \qquad\qquad (53) \\
&\quad - \sum_{k=2}^{\infty} G_{2k}(\omega_1, \omega_2) x^{2k-1} \\
&= x^{-1} - \lim \sum (m\omega_2 + n\omega_1)^{-2} x \\
&\quad - \sum_{k=2}^{\infty} \sum_{(m,n)\neq(0,0)} (m\omega_2 + n\omega_1)^{-2k} x^{2k-1} \\
&= x^{-1} - \lim \sum (m\omega_2 + n\omega_1)^{-2} x \\
&\quad - \sum_{(m,n)\neq(0,0)} \sum_{k=2}^{\infty} (m\omega_2 + n\omega_1)^{-2k} x^{2k-1}.
\end{aligned}
$$

The interchange of the sums is justified by absolute convergence. It was to arrange this absolute convergence that we split off the $G_2(\omega_1, \omega_2)x$ term from the rest of the sum. The order in which we take the outer sum in the last term on the right is irrelevant. We choose to take the same order as in the previous term.

$$\begin{aligned}
\phi(x, \omega_1, \omega_2) &= x^{-1} - \lim \sum (m\omega_2 + n\omega_1)^{-2} x \qquad (54) \\
&\quad - \lim \sum_{k=2}^{\infty} \sum (m\omega_2 + n\omega_1)^{-2k} x^{2k-1} \\
&= x^{-1} - \lim \sum_{\substack{-M \le m \le M \\ -N \le n \le N \\ (m, n) \ne (0, 0)}} \sum_{k=1}^{\infty} (m\omega_2 + n\omega_1)^{-2k} x^{2k-1} \\
&= x^{-1} + \lim \sum \frac{1}{2}[(x - m\omega_2 - n\omega_2)^{-1} \\
&\quad + (x + m\omega_2 + n\omega_2)^{-1}]
\end{aligned}$$

or, using the symmetry of the sum,

$$\begin{aligned}
\phi(x, \omega_1, \omega_2) &= x^{-1} + \lim \sum (x - m\omega_2 - n\omega_2)^{-1} \qquad (55) \\
&= \lim \sum_{\substack{-M \le m \le M \\ -N \le n \le N}} (x - m\omega_2 - n\omega_2)^{-1}.
\end{aligned}$$

Note that, except in the last line of (53), where we could use absolute convergence, we have interchanged limits only with *finite* sums.

The corresponding expansion for f_1 is

$$f_1(\tau, \zeta) = \lim \sum_{\substack{-M \le m \le M \\ -N \le n \le N}} (\zeta - m\tau - n)^{-1}, \qquad (56)$$

or, using equation (42),

$$f_1(\tau, \zeta) = \lim_{M \to \infty} \sum_{-M \le m \le M} \pi \cot \pi(\zeta - m\tau). \qquad (57)$$

Integrating and then exponentiating, we find the product expansion

$$\theta_1(\tau, \zeta) = \frac{1}{\pi}\theta_1'(\tau, 0) \sin \pi\zeta \lim_{M \to \infty} \prod_{\substack{-M \le m \le M \\ m \ne 0}} \frac{\sin \pi(\zeta - m\tau)}{\sin \pi(-m\tau)} \tag{58}$$

$$= \frac{1}{\pi}\theta_1'(\tau, 0) \sin \pi\zeta \lim_{M \to \infty} \prod_{m=1}^{M} \frac{\sin \pi(\zeta - m\tau)}{\sin \pi(-m\tau)} \frac{\sin \pi(\zeta + m\tau)}{\sin \pi(m\tau)}$$

$$= C\frac{z^{1/2} - z^{-1/2}}{2i} \prod_{m=1}^{\infty} \frac{(1 - q^{2m}z^{-1})(1 - q^{2m}z)}{(1 - q^{2m})(1 - q^{2m})}.$$

with the "constant"

$$C = \frac{1}{\pi}\theta_1'(\tau, 0). \tag{59}$$

The quotation marks on the word constant reflect the fact that C is a function of τ.

We can derive product expansions for the other theta functions from the product representation (58) for θ_1 and the translation relations. These are

$$\theta_0(\tau, \zeta) = Cq^{-1/4} \prod_{m=1}^{\infty}(1 - q^{-2m})^{-2}(1 - q^{2m-1}z^{-1})(1 - q^{2m-1}z), \tag{60}$$

$$\theta_2(\tau, \zeta) = C\frac{z^{1/2} + z^{-1/2}}{2} \prod_{m=1}^{\infty}(1 - q^{-2m})^{-2}(1 + q^{2m}z^{-1})(1 + q^{2m}z),$$

$$\theta_3(\tau, \zeta) = Cq^{-1/4} \prod_{m=1}^{\infty}(1 - q^{-2m})^{-2}(1 + q^{2m-1}z^{-1})(1 + q^{2m-1}z).$$

We will be able to write these more simply once we prove the Jacobi identity in the next section. For now we note that

$$\theta_0(\tau, 0)\theta_2(\tau, 0)\theta_3(\tau, 0) = C^3q^{-1/2} \prod_{m=1}^{\infty}(1 - q^{-2m})^{-6}. \tag{61}$$

Exercises

1. Find the Laurent expansions for the logarithmic derivatives of θ_0, θ_2, and θ_3.

2. Find transformation formulae analogous to (51) for θ_0, θ_2, and θ_3. These can be found by the same method used for θ_1, or they can be derived from (51) and the translation relations.

3. Prove that

$$\theta_1(\tau,\zeta) = \frac{\partial\theta_1}{\partial\zeta}(\tau,0)\lim_{M\to\infty}\lim_{N\to\infty}\zeta\prod_{\substack{-M\le m\le M\\-N\le n\le N\\(m,n)\ne(0,0)}}(1-\frac{\zeta}{m\tau+n}).$$

3.3 Lattices

A *lattice*[5] is a discrete subgroup of a finite dimensional Euclidean space. We can choose a basis for this Euclidean space such that the inner product takes the usual form $< x, y >= \sum_{j=1}^{n} x_j y_j$. There is a deep connection between the properties of theta functions and those of lattices. Lattices in low dimensions are well understood and provide information about theta functions. In higher dimensions the situation is reversed, with theta functions providing information about lattices. Both techniques are due to Jacobi and the present section is essentially copied from Borchardt's notes of Jacobi's lectures, printed in [4].

If L is a lattice in $\mathbf{R}^n$ and χ a function on L then we define

$$\Theta_{L,\chi}(\tau,\zeta) = \sum_{\lambda\in L}\chi(\lambda)e^{\pi i<\lambda,\lambda>\tau+2\pi i<\lambda,\zeta>} \tag{1}$$

for Im $\tau > 0$ and $\zeta \in \mathbf{C}^n$. Although this definition makes sense for fairly arbitrary functions χ, we will be interested only in the case where χ is constant on cosets[6] of some sublattice L' of finite index in L. In this case the sum certainly converges absolutely, so we need not specify the order of summation. The sum is locally uniformly convergent, and thus an analytic function.

The classical θ_j's are all of this type. If $L = \frac{1}{2}\mathbf{Z}$, $L' = 2\mathbf{Z}$, and functions χ_j are defined by giving their values on coset representatives,

$$\begin{array}{llll}
\chi_0(0)=1 & \chi_0(\tfrac{1}{2})=0 & \chi_0(1)=-1 & \chi_0(\tfrac{3}{2})=0\\
\chi_1(0)=0 & \chi_1(\tfrac{1}{2})=-i & \chi_1(1)=0 & \chi_1(\tfrac{3}{2})=i\\
\chi_2(0)=0 & \chi_2(\tfrac{1}{2})=1 & \chi_2(1)=0 & \chi_2(\tfrac{3}{2})=1\\
\chi_3(0)=1 & \chi_3(\tfrac{1}{2})=0 & \chi_3(1)=1 & \chi_3(\tfrac{3}{2})=0,
\end{array} \tag{2}$$

then

$$\Theta_{L,\chi_j}(\tau,\zeta) = \theta_j(\tau,\zeta). \tag{3}$$

---

[5]The word lattice has various meanings in various branches of mathematics. It has at least two meanings within the theory of theta functions. Often a discrete subgroup of $\mathbf{C}$ with compact quotient is called a lattice. To avoid confusion we will not use the word lattice in this sense.

[6]A *coset* of a sublattice L' of a lattice L is a set of the form $\lambda + L'$ where $\lambda \in L$. A set $R = \{\lambda_1,\dots,\lambda_n\}$ of elements of L is called a *set of coset representatives* if every element $\lambda \in L$ can be written in the form $\lambda = \lambda_j + \lambda'$ with a unique $\lambda_j \in S$ and $\lambda' \in L'$. The number n is called the *index* of L' in L.

Since $\chi_0, \chi_1, \chi_2, \chi_3$ form a basis for the functions on L/L', it is clear that any $\Theta_{L,\chi}$ with this L and L' can be written as a linear combination of $\theta_0, \theta_1, \theta_2, \theta_3$.

There is very little interesting to be said about lattices in one dimension. Up to dilation they are all the lattice $\mathbf{Z}$. Like all lattices, they are invariant under multiplication by -1. This latter property is helpful in determining parity relations. We compute

$$\Theta_{L,\chi}(\tau, -\zeta) = \sum_{\lambda \in L} \chi(\lambda) e^{\pi i <\lambda,\lambda> \tau + 2\pi i <\lambda, -\zeta>} \tag{4}$$

$$= \sum_{-\lambda \in L} \tilde{\chi}(-\lambda) e^{\pi i <-\lambda,-\lambda> \tau + 2\pi i <-\lambda, \zeta>}$$

$$= \Theta_{L,\tilde{\chi}}(\tau, \zeta)$$

where $\tilde{\chi}(\lambda) = \chi(-\lambda)$. Thus $\Theta_{L,\chi}$ is even if χ is even and odd if χ is odd. Of the functions χ_j listed above, χ_1 is odd, while the others are all even. It follows that θ_1 is odd while the other theta functions are even, just as we found in (5) of the preceding section.

The argument above has two important generalizations. Suppose that A is an orthogonal transformation, so that

$$< Ax, Ay >=< x, y > . \tag{5}$$

Then

$$\Theta_{L,\chi}(\tau, A\zeta) = \sum_{\lambda \in L} \chi(\lambda) e^{\pi i <\lambda,\lambda> \tau + 2\pi i <\lambda, A\zeta>} \tag{6}$$

$$= \sum_{A^{-1}\lambda \in A^{-1}L} (A^{-1}\chi)(A^{-1}\lambda) e^{\pi i <A^{-1}\lambda, A^{-1}\lambda> \tau + 2\pi i <A^{-1}\lambda, \zeta>}$$

$$= \Theta_{A^{-1}L, A^{-1}\chi}(\tau, \zeta)$$

where $A^{-1}\chi(\lambda) = \chi(A\lambda)$ is a function on $A^{-1}L$ which is constant on cosets of $A^{-1}L'$. The most interesting special case occurs when A preserves L and L', that is, when $AL = L$ and $AL' = L'$. Of course the only orthogonal transformations on $\mathbf{R}$ are ± 1, so we will get nothing new until we move to higher dimensions.

The other interesting generalization is to dilations by a factor of $\alpha \in \mathbf{Q} - \{0\}$,

$$\Theta_{L,\chi}(\tau, \alpha\zeta) = \sum_{\lambda \in L} \chi(\lambda) e^{\pi i <\lambda,\lambda> \tau + 2\pi i <\lambda, \alpha\zeta>} \tag{7}$$

$$= \sum_{\alpha\lambda \in \alpha L} \tilde{\chi}(\alpha\lambda) e^{\pi i <\alpha\lambda, \alpha\lambda> \alpha^{-2}\tau + 2\pi i <\alpha\lambda, \zeta>}$$

$$= \Theta_{\alpha L, \tilde{\chi}}(\alpha^{-2}\tau, \zeta),$$

or, more conveniently,

$$\Theta_{L,\chi}(\alpha^2\tau, \zeta) = \Theta_{\alpha L, \tilde{\chi}}(\tau, \alpha^{-1}\zeta) \tag{8}$$

where $\tilde{\chi}(\lambda) = \chi(\alpha^{-1}\lambda)$. The most interesting special case occurs when α is an integer and χ is constant on cosets of $\alpha^{-1}L'$. In this case $\tilde{\chi}$ is constant on cosets of L', so we have another theta function of the same type. For example if $L = \frac{1}{2}\mathbf{Z}$, $L' = 2\mathbf{Z}$, and $n = 2$ then both χ_2 and χ_3 are constant on cosets of $\alpha^{-1}L = \mathbf{Z}$. If $\chi = \chi_2$ then $\tilde{\chi} = \frac{1}{2}\chi_3 - \frac{1}{2}\chi_0$ so

$$\theta_2(4\tau, \zeta) = \frac{1}{2}\theta_3(\tau, \frac{1}{2}\zeta) - \frac{1}{2}\theta_0(\tau, \frac{1}{2}\zeta). \tag{9}$$

If $\chi = \chi_3$ then $\tilde{\chi} = \frac{1}{2}\chi_3 + \frac{1}{2}\chi_0$ so

$$\theta_3(4\tau, \zeta) = \frac{1}{2}\theta_3(\tau, \frac{1}{2}\zeta) + \frac{1}{2}\theta_0(\tau, \frac{1}{2}\zeta). \tag{10}$$

We could, of course, have proved the relations (9) and (10) directly from the infinite sums defining the theta functions, without introducing the concept of lattices. So far we have gained no essentially new information.

　　To proceed further we must consider lattices in spaces of dimension greater than one. We define lattices $L_m \in \mathbf{R}^4$ by

$$L_m = \{\lambda \in \mathbf{Z}[\frac{1}{2}]^4 : \; <\lambda, \lambda > \in 2^m\mathbf{Z}\}, \tag{11}$$

where $\mathbf{Z}[\frac{1}{2}]$ is the ring of all rational numbers whose denominator is a power of 2. It is clear that

$$2^m\mathbf{Z}^4 \subset L_m \subset 2^{m-1}\mathbf{Z}^4, \tag{12}$$

so L_m is really a lattice. We define matrices A and D by

$$A = -\frac{1}{2}\begin{pmatrix} +1 & -1 & -1 & -1 \\ +1 & +1 & -1 & +1 \\ +1 & +1 & +1 & -1 \\ +1 & -1 & +1 & +1 \end{pmatrix} \tag{13}$$

and

$$D = \begin{pmatrix} +1 & -1 & 0 & 0 \\ +1 & +1 & 0 & 0 \\ 0 & 0 & +1 & -1 \\ 0 & 0 & +1 & +1 \end{pmatrix}. \tag{14}$$

It is easy to check that $A^T A = I$ and $D^T D = 2I$. From this we draw two conclusions. First, A, A^{-1}, D, and D^{-1} all have entries in $\mathbf{Z}[\frac{1}{2}]$, so the statements $\lambda \in \mathbf{Z}[\frac{1}{2}]^4$, $A\lambda \in \mathbf{Z}[\frac{1}{2}]^4$, and $D\lambda \in \mathbf{Z}[\frac{1}{2}]^4$ are all equivalent. Second, the statements $< \lambda, \lambda > \in 2^m$, $< A\lambda, A\lambda > \in 2^m$, and $< D\lambda, D\lambda > \in 2^{m+1}$ are all equivalent. In view of the definition of L_m, these two facts imply

$$AL_m = L_m \tag{15}$$

and

$$DL_m = L_{m+1}. \tag{16}$$

Since $\det(D) = 4$ we see immediately that the index of L_{m+1} in L_m is 4.

Initially we will take $L = L_0$ and $L' = L_1$ in (1). We begin by choosing coset representatives for L/L',

$$\alpha = \begin{pmatrix} 0 \\ 0 \\ 0 \\ 0 \end{pmatrix}, \quad \beta = \begin{pmatrix} 1 \\ 0 \\ 0 \\ 0 \end{pmatrix}, \quad \gamma = -\frac{1}{2}\begin{pmatrix} 1 \\ 1 \\ 1 \\ 1 \end{pmatrix}, \quad \delta = -\frac{1}{2}\begin{pmatrix} 1 \\ -1 \\ -1 \\ -1 \end{pmatrix} \tag{17}$$

The action of A on the coset representatives is

$$A\alpha = \alpha, \quad A\beta = \gamma, \quad A\gamma = \delta, \quad A\delta = \beta. \tag{18}$$

We choose the following basis for functions on L/L':

$$\begin{array}{llll}
\chi_{0000}(\alpha) = 1 & \chi_{0000}(\beta) = -1 & \chi_{0000}(\gamma) = 0 & \chi_{0000}(\delta) = 0 \\
\chi_{1111}(\alpha) = 0 & \chi_{1111}(\beta) = 0 & \chi_{1111}(\gamma) = 1 & \chi_{1111}(\delta) = -1 \\
\chi_{2222}(\alpha) = 0 & \chi_{2222}(\beta) = 0 & \chi_{2222}(\gamma) = 1 & \chi_{2222}(\delta) = 1 \\
\chi_{3333}(\alpha) = 1 & \chi_{3333}(\beta) = 1 & \chi_{3333}(\gamma) = 0 & \chi_{3333}(\delta) = 0.
\end{array} \tag{19}$$

These have been chosen in such a way that

$$\chi_{jjjj}(\lambda) = \chi_j(\lambda_1)\chi_j(\lambda_2)\chi_j(\lambda_2)\chi_j(\lambda_4) \tag{20}$$

where χ_j is as in (2). We can easily compute the action of A^{-1} on the χ's from the action on the coset representatives,

$$A^{-1}\begin{pmatrix} \chi_{0000} \\ \chi_{1111} \\ \chi_{2222} \\ \chi_{3333} \end{pmatrix} = \frac{1}{2}\begin{pmatrix} +1 & +1 & -1 & +1 \\ -1 & -1 & -1 & +1 \\ -1 & +1 & +1 & +1 \\ +1 & -1 & +1 & +1 \end{pmatrix}\begin{pmatrix} \chi_{0000} \\ \chi_{1111} \\ \chi_{2222} \\ \chi_{3333} \end{pmatrix}. \tag{21}$$

Equation (6) then shows that

$$\begin{pmatrix} \Theta_{0000} \\ \Theta_{1111} \\ \Theta_{2222} \\ \Theta_{3333} \end{pmatrix} (\tau, A\zeta) = \frac{1}{2} \begin{pmatrix} +1 & +1 & -1 & +1 \\ -1 & -1 & -1 & +1 \\ -1 & +1 & +1 & +1 \\ +1 & -1 & +1 & +1 \end{pmatrix} \begin{pmatrix} \Theta_{0000} \\ \Theta_{1111} \\ \Theta_{2222} \\ \Theta_{3333} \end{pmatrix} (\tau, \zeta), \qquad (22)$$

where we have written

$$\Theta_{jjjj}$$

in place of

$$\Theta_{L_0, \chi_{jjjj}}.$$

These formulae would have a nicer appearance if we had chosen a different basis of functions on L/L'. The advantage of the choice we have made is that the Θ's are easily expressed in terms of the θ's,

$$\Theta_{jjjj}(\tau, (\zeta_1, \zeta_2, \zeta_3, \zeta_4)^T) = \theta_j(\tau, \zeta_1)\theta_j(\tau, \zeta_2)\theta_j(\tau, \zeta_3)\theta_j(\tau, \zeta_4). \qquad (23)$$

This follows from (20). We simply extend χ_{jjjj} to a function on $\frac{1}{2}\mathbf{Z}^4$ by setting $\chi_{jjjj}(\lambda) = 0$ if $\lambda \notin L_0$. This does not change the sum (1), so

$$\begin{aligned} \Theta_{jjjj}(\tau, \zeta) &= \sum_{\lambda \in \frac{1}{2}\mathbf{Z}^4} \chi_{jjjj}(\lambda) e^{\pi i <\lambda, \lambda> \tau + 2\pi i <\lambda, \zeta>} \qquad (24) \\[2mm] &= \sum_{\lambda_1, \lambda_2, \lambda_3, \lambda_4 \in \frac{1}{2}\mathbf{Z}} \prod_{k=1}^{4} \chi_j(\lambda_k) e^{\pi i \sum_{k=1}^{4} \lambda_k^2 \tau + 2\pi i \sum_{k=1}^{4} \lambda_k \zeta_k} \\[2mm] &= \prod_{k=1}^{4} \sum_{\lambda_k \in \frac{1}{2}\mathbf{Z}} \chi_j(\lambda_k) e^{\pi i \lambda_k^2 \tau + 2\pi i \lambda_k \zeta_k} \\[2mm] &= \prod_{k=1}^{4} \theta_j(\tau, \zeta_k). \end{aligned}$$

One rarely needs equation (22) in its full generality. Normally one specializes ζ in some way. For example, if we take $\zeta = \xi\delta$ with $\xi \in \mathbf{C}$ then $A\zeta = \xi\beta$ while

$$\begin{pmatrix} \theta_{0000} \\ \theta_{1111} \\ \theta_{2222} \\ \theta_{3333} \end{pmatrix} (\tau, \xi\beta) = \begin{pmatrix} \theta_0(\tau, 0)^3 \theta_0(\tau, \xi) \\ \theta_1(\tau, 0)^3 \theta_1(\tau, \xi) \\ \theta_2(\tau, 0)^3 \theta_2(\tau, \xi) \\ \theta_3(\tau, 0)^3 \theta_3(\tau, \xi) \end{pmatrix} \qquad (25)$$

and

$$\begin{pmatrix} \theta_{0000} \\ \theta_{1111} \\ \theta_{2222} \\ \theta_{3333} \end{pmatrix} (\tau, \xi\delta) = \begin{pmatrix} \theta_0(\tau, \frac{1}{2}\xi)^4 \\ -\theta_1(\tau, \frac{1}{2}\xi)^4 \\ \theta_2(\tau, \frac{1}{2}\xi)^4 \\ \theta_3(\tau, \frac{1}{2}\xi)^4 \end{pmatrix}. \tag{26}$$

Equation (21) then gives

$$\begin{pmatrix} \theta_0(\tau, 0)^3\theta_0(\tau, \xi) \\ \theta_1(\tau, 0)^3\theta_1(\tau, \xi) \\ \theta_2(\tau, 0)^3\theta_2(\tau, \xi) \\ \theta_3(\tau, 0)^3\theta_3(\tau, \xi) \end{pmatrix} = \frac{1}{2}\begin{pmatrix} +1 & -1 & -1 & +1 \\ -1 & +1 & -1 & +1 \\ -1 & -1 & +1 & +1 \\ +1 & +1 & +1 & +1 \end{pmatrix} \begin{pmatrix} \theta_0(\tau, \frac{1}{2}\xi)^4 \\ \theta_1(\tau, \frac{1}{2}\xi)^4 \\ \theta_2(\tau, \frac{1}{2}\xi)^4 \\ \theta_3(\tau, \frac{1}{2}\xi)^4 \end{pmatrix}. \tag{27}$$

Since $\theta_1(\tau, 0) = 0$ the second component of the vector on the left, and hence on the right as well, vanishes for all $\xi \in \mathbf{C}$. This gives the relation

$$\theta_0(\tau, \xi)^4 + \theta_2(\tau, \xi)^4 = \theta_1(\tau, \xi)^4 + \theta_3(\tau, \xi)^4. \tag{28}$$

An important special case occurs when $\xi = 0$,

$$\theta_3(\tau, 0)^4 = \theta_0(\tau, 0)^4 + \theta_2(\tau, 0)^4. \tag{29}$$

We can use (28) to simplify the other three relations in (27). Thus, instead of

$$\theta_2(\tau, 0)^3\theta_2(\tau, \xi) = \frac{1}{2}[\theta_3(\tau, \frac{1}{2}\xi)^4 + \theta_2(\tau, \frac{1}{2}\xi)^4 - \theta_1(\tau, \frac{1}{2}\xi)^4 - \theta_0(\tau, \frac{1}{2}\xi)^4], \tag{30}$$

we can write

$$\theta_2(\tau, 0)^3\theta_2(\tau, \xi) = \theta_3(\tau, \frac{1}{2}\xi)^4 - \theta_0(\tau, \frac{1}{2}\xi)^4, \tag{31}$$

or

$$\theta_2(\tau, 0)^3\theta_2(\tau, \xi) = \theta_2(\tau, \frac{1}{2}\xi)^4 - \theta_1(\tau, \frac{1}{2}\xi)^4. \tag{32}$$

We now have enough identities to prove Jacobi's triple product identity. Taking (31) and factoring the right hand side we find

$$\theta_2(\tau, 0)^3\theta_2(\tau, \xi) = [\theta_3(\tau, \frac{1}{2}\xi) - \theta_0(\tau, \frac{1}{2}\xi)][\theta_3(\tau, \frac{1}{2}\xi) + \theta_0(\tau, \frac{1}{2}\xi)] \tag{33}$$

$$\times [\theta_3(\tau, \frac{1}{2}\xi)^2 + \theta_0(\tau, \frac{1}{2}\xi)^2].$$

We use equations (9) and (10) to simplify the right hand side

$$\theta_2(\tau, 0)^3 \theta_2(\tau, \xi) = 4\theta_2(4\tau, \xi)\theta_3(4\tau, \xi)[\theta_3(\tau, \tfrac{1}{2}\xi)^2 + \theta_0(\tau, \tfrac{1}{2}\xi)^2]. \qquad (34)$$

Writing the last factor as

$$\frac{1}{2}[\theta_3(\tau, \tfrac{1}{2}\xi) - \theta_0(\tau, \tfrac{1}{2}\xi)]^2 + \frac{1}{2}[\theta_3(\tau, \tfrac{1}{2}\xi) + \theta_0(\tau, \tfrac{1}{2}\xi)]^2$$

and using (9) and (10) again,

$$\theta_2(\tau, 0)^3 \theta_2(\tau, \xi) = 8\theta_2(4\tau, \xi)\theta_3(4\tau, \xi)[\theta_2(4\tau, \xi)^2 + \theta_3(4\tau, \xi)^2]. \qquad (35)$$

Letting $\xi = \eta + \tfrac{1}{2}$ and using the translation relations (3),

$$\theta_2(\tau, 0)^3 \theta_1(\tau, \eta) = 8\theta_0(4\tau, \eta)\theta_1(4\tau, \eta)[\theta_0(4\tau, \eta)^2 + \theta_1(4\tau, \eta)^2]. \qquad (36)$$

We now differentiate with respect to η and set $\eta = 0$. In this way we obtain the identity

$$\theta_2(\tau, 0)^3 \theta_1'(\tau, 0) = 8\theta_0(4\tau, 0)^3 \theta_1'(4\tau, 0). \qquad (37)$$

Applying (35) to $\xi = 0$ we find

$$\theta_2(\tau, 0)^4 = 8\theta_0(4\tau, 0)\theta_2(4\tau, 0)[\theta_0(4\tau, 0)^2 + \theta_2(4\tau, 0)^2]. \qquad (38)$$

We can solve equations (9) and (10) for $\theta_0(\tau, \tfrac{1}{2}\zeta)$ and $\theta_3(\tau, \tfrac{1}{2}\zeta)$,

$$\theta_0(\tau, \tfrac{1}{2}\zeta) = \theta_3(4\tau, \zeta) - \theta_2(4\tau, \zeta), \qquad (39)$$

$$\theta_3(\tau, \tfrac{1}{2}\zeta) = \theta_3(4\tau, \zeta) + \theta_2(4\tau, \zeta).$$

Setting $\zeta = 0$ and multiplying these equations gives

$$\theta_0(\tau, 0)\theta_3(\tau, 0) = \theta_3(4\tau, 0)^2 - \theta_2(4\tau, 0)^2. \qquad (40)$$

Multiplying (38) and (40),

$$\begin{aligned}
\theta_0(\tau, 0)\theta_2(\tau, 0)^4 \theta_3(\tau, 0) &= 8\theta_2(4\tau, 0)\theta_3(4\tau, 0) \\
&\quad \times [\theta_3(4\tau, 0)^4 - \theta_2(4\tau, 0)^4].
\end{aligned} \qquad (41)$$

By (29) this is the same as

$$\theta_0(\tau, 0)\theta_2(\tau, 0)^4 \theta_3(\tau, 0) = 8\theta_0(4\tau, 0)^4 \theta_2(4\tau, 0)\theta_3(4\tau, 0). \qquad (42)$$

Dividing (37) by (42),

$$\frac{\theta_1'(\tau, 0)}{\theta_0(\tau, 0)\theta_2(\tau, 0)\theta_3(\tau, 0)} = \frac{\theta_1'(4\tau, 0)}{\theta_0(4\tau, 0)\theta_2(4\tau, 0)\theta_3(4\tau, 0)}. \tag{43}$$

Thus the function

$$\Xi(\tau) = \frac{\theta_1'(\tau, 0)}{\theta_0(\tau, 0)\theta_2(\tau, 0)\theta_3(\tau, 0)} \tag{44}$$

is unchanged when τ is multiplied by 4. It follows that

$$\Xi(\tau) = \lim_{n \to \infty} \Xi(4^n \tau) = \pi. \tag{45}$$

The limit is evaluated by writing Ξ in terms of q and taking the limit as q tends to 0. From (45) we get the Jacobi identity

$$\theta_1'(\tau, 0) = \pi \theta_0(\tau, 0)\theta_2(\tau, 0)\theta_3(\tau, 0). \tag{46}$$

The main use of the Jacobi identity is in evaluating "constants" which depend on τ. As an example we use it here to compute the constants in the product representations (58) and (60). It can also be used to put equation (51) of Section 3.2 in a more useful form, but we will leave this as an exercise.

THEOREM 30 (PRODUCT REPRESENTATIONS OF THETA FUNCTIONS)
The theta functions have the following product representations:

$$\theta_0(\tau, \zeta) = \prod_{m=1}^{\infty} (1 - q^{2m-1}z^{-1})(1 - q^{2m})(1 - q^{2m-1}z), \tag{47}$$

$$\theta_1(\tau, \zeta) = \frac{z^{1/2} - z^{-1/2}}{2i} q^{1/4} \prod_{m=1}^{\infty} (1 - q^{2m}z^{-1})(1 - q^{2m})(1 - q^{2m}z),$$

$$\theta_2(\tau, \zeta) = \frac{z^{1/2} + z^{-1/2}}{2} q^{1/4} \prod_{m=1}^{\infty} (1 + q^{2m}z^{-1})(1 - q^{2m})(1 + q^{2m}z),$$

$$\theta_3(\tau, \zeta) = \prod_{m=1}^{\infty} (1 + q^{2m-1}z^{-1})(1 - q^{2m})(1 + q^{2m-1}z).$$

Proof. First we note that (61) and (46) imply

$$\theta_1'(\tau, 0) = \pi^{-2}\theta_1'(\tau, 0)^3 q^{-1/2} \prod_{m=1}^{\infty} (1 - q^{2m})^{-6} \tag{48}$$

or

$$\theta_1'(\tau, 0)^2 = \pi^2 q^{1/2} \prod_{m=1}^{\infty} (1 - q^{2m})^6. \tag{49}$$

From the definition (1) of θ_1 we see that

$$\lim_{\tau \to i\infty} e^{-\pi i \tau/4} \theta_1'(\tau, 0) = \pi, \tag{50}$$

so the correct choice of square root is

$$\theta_1'(\tau, 0) = \pi q^{1/4} \prod_{m=1}^{\infty} (1 - q^{2m})^3. \tag{51}$$

Inserting (51) in (58) and (60) yields (47).

The 4 identities in (21) are part of a system of 64 identities. For any 4-tuple j, k, l, m with $1 \le j, k, l, m \le 4$, we can define a function χ_{jklm} by

$$\chi_{jklm}(\lambda) = \chi_j(\lambda_1) \chi_k(\lambda_2) \chi_l(\lambda_2) \chi_m(\lambda_4). \tag{52}$$

There are $4^4 = 256$ such functions. In general χ_{jklm} is zero off of $\frac{1}{2}\mathbf{Z}^4$ and is invariant under translations by elements of $2\mathbf{Z}^4$. There are, however 64 cases in which χ_{jklm} is zero off of L_{-1}, and is invariant under translations by elements of L_2. Four of these we have already met. When $j = k = l = m$ more is true: χ_{jklm} is zero off of L_0 and is invariant under translations by elements of L_1. An additional 24 cases occur when $jklm$ is a permutation of 1234. These can be divided into two groups of 12. The first group is zero off of L_{-1} and is invariant under translations by elements of L_1, while the second group is zero off of L_0 and is invariant under translations by elements of L_2. Finally there are 36 cases where j is equal to precisely one of $k, jl,$ or m and the remaining two are equal to each other. This group is zero off of L_{-1} and is invariant under translations by elements of L_2. We will write

$$\Theta_{jklm}$$

in place of

$$\Theta_{L_{-1}, \chi_{jklm}}.$$

We can now proceed as before to find 64 identities. Each of these expresses some theta function, $\Theta_{jklm}(\tau, A\zeta)$, as a linear combination of four other theta functions, $\Theta_{j'k'l'm'}(\tau, \zeta)$, with various 4-tuples j', k', l', m'. Four of these will, of course, be the identities (22) which we already possess. The remaining 60 are

$$\begin{pmatrix} \Theta_{2301} \\ \Theta_{3210} \\ \Theta_{0123} \\ \Theta_{1032} \end{pmatrix}(\tau, A\zeta) = \frac{1}{2}\begin{pmatrix} +1 & -1 & -1 & -1 \\ -1 & +1 & -1 & -1 \\ -1 & -1 & +1 & -1 \\ +1 & +1 & +1 & -1 \end{pmatrix}\begin{pmatrix} \Theta_{2130} \\ \Theta_{3021} \\ \Theta_{0312} \\ \Theta_{1203} \end{pmatrix}(\tau, \zeta), \qquad (53)$$

$$\begin{pmatrix} \Theta_{2130} \\ \Theta_{3021} \\ \Theta_{0312} \\ \Theta_{1203} \end{pmatrix}(\tau, A\zeta) = \frac{1}{2}\begin{pmatrix} +1 & -1 & -1 & -1 \\ -1 & +1 & -1 & -1 \\ -1 & -1 & +1 & -1 \\ +1 & +1 & +1 & -1 \end{pmatrix}\begin{pmatrix} \Theta_{2013} \\ \Theta_{3102} \\ \Theta_{0231} \\ \Theta_{1320} \end{pmatrix}(\tau, \zeta), \qquad (54)$$

$$\begin{pmatrix} \Theta_{2013} \\ \Theta_{3102} \\ \Theta_{0231} \\ \Theta_{1320} \end{pmatrix}(\tau, A\zeta) = \frac{1}{2}\begin{pmatrix} +1 & -1 & -1 & -1 \\ -1 & +1 & -1 & -1 \\ -1 & -1 & +1 & -1 \\ +1 & +1 & +1 & -1 \end{pmatrix}\begin{pmatrix} \Theta_{2301} \\ \Theta_{3210} \\ \Theta_{0123} \\ \Theta_{1032} \end{pmatrix}(\tau, \zeta), \qquad (55)$$

$$\begin{pmatrix} \Theta_{2310} \\ \Theta_{3201} \\ \Theta_{0132} \\ \Theta_{1023} \end{pmatrix}(\tau, A\zeta) = \frac{1}{2}\begin{pmatrix} +1 & -1 & -1 & -1 \\ -1 & +1 & -1 & -1 \\ -1 & -1 & +1 & -1 \\ +1 & +1 & +1 & -1 \end{pmatrix}\begin{pmatrix} \Theta_{2031} \\ \Theta_{3120} \\ \Theta_{0213} \\ \Theta_{1302} \end{pmatrix}(\tau, \zeta), \qquad (56)$$

$$\begin{pmatrix} \Theta_{2031} \\ \Theta_{3120} \\ \Theta_{0213} \\ \Theta_{1302} \end{pmatrix}(\tau, A\zeta) = \frac{1}{2}\begin{pmatrix} +1 & -1 & -1 & -1 \\ -1 & +1 & -1 & -1 \\ -1 & -1 & +1 & -1 \\ +1 & +1 & +1 & -1 \end{pmatrix}\begin{pmatrix} \Theta_{2103} \\ \Theta_{3021} \\ \Theta_{0321} \\ \Theta_{1230} \end{pmatrix}(\tau, \zeta), \qquad (57)$$

$$\begin{pmatrix} \Theta_{2103} \\ \Theta_{3021} \\ \Theta_{0321} \\ \Theta_{1230} \end{pmatrix}(\tau, A\zeta) = \frac{1}{2}\begin{pmatrix} +1 & -1 & -1 & -1 \\ -1 & +1 & -1 & -1 \\ -1 & -1 & +1 & -1 \\ +1 & +1 & +1 & -1 \end{pmatrix}\begin{pmatrix} \Theta_{2310} \\ \Theta_{3201} \\ \Theta_{0213} \\ \Theta_{1302} \end{pmatrix}(\tau, \zeta), \qquad (58)$$

$$\begin{pmatrix} \Theta_{1100} \\ \Theta_{0011} \\ \Theta_{3322} \\ \Theta_{2233} \end{pmatrix}(\tau, A\zeta) = \frac{1}{2}\begin{pmatrix} -1 & -1 & +1 & -1 \\ +1 & +1 & +1 & -1 \\ -1 & +1 & +1 & +1 \\ +1 & -1 & +1 & +1 \end{pmatrix}\begin{pmatrix} \Theta_{1010} \\ \Theta_{0101} \\ \Theta_{3232} \\ \Theta_{2323} \end{pmatrix}(\tau, \zeta), \qquad (59)$$

$$\begin{pmatrix} \Theta_{1010} \\ \Theta_{0101} \\ \Theta_{3232} \\ \Theta_{2323} \end{pmatrix}(\tau, A\zeta) = \frac{1}{2}\begin{pmatrix} -1 & -1 & +1 & -1 \\ +1 & +1 & +1 & -1 \\ -1 & +1 & +1 & +1 \\ +1 & -1 & +1 & +1 \end{pmatrix}\begin{pmatrix} \Theta_{1001} \\ \Theta_{0110} \\ \Theta_{3223} \\ \Theta_{2332} \end{pmatrix}(\tau, \zeta), \qquad (60)$$

$$\begin{pmatrix} \Theta_{1001} \\ \Theta_{0110} \\ \Theta_{3223} \\ \Theta_{2332} \end{pmatrix}(\tau, A\zeta) = \frac{1}{2}\begin{pmatrix} -1 & -1 & +1 & -1 \\ +1 & +1 & +1 & -1 \\ -1 & +1 & +1 & +1 \\ +1 & -1 & +1 & +1 \end{pmatrix}\begin{pmatrix} \Theta_{1100} \\ \Theta_{0011} \\ \Theta_{3322} \\ \Theta_{2323} \end{pmatrix}(\tau, \zeta), \qquad (61)$$

$$\begin{pmatrix} \Theta_{2200} \\ \Theta_{3311} \\ \Theta_{0022} \\ \Theta_{1133} \end{pmatrix}(\tau, A\zeta) = \frac{1}{2}\begin{pmatrix} +1 & -1 & +1 & +1 \\ -1 & +1 & +1 & +1 \\ +1 & +1 & +1 & -1 \\ -1 & -1 & +1 & -1 \end{pmatrix}\begin{pmatrix} \Theta_{2020} \\ \Theta_{3131} \\ \Theta_{0202} \\ \Theta_{1313} \end{pmatrix}(\tau, \zeta), \qquad (62)$$

$$\begin{pmatrix} \Theta_{2020} \\ \Theta_{3131} \\ \Theta_{0202} \\ \Theta_{1313} \end{pmatrix}(\tau, A\zeta) = \frac{1}{2}\begin{pmatrix} +1 & -1 & +1 & +1 \\ -1 & +1 & +1 & +1 \\ +1 & +1 & +1 & -1 \\ -1 & -1 & +1 & -1 \end{pmatrix}\begin{pmatrix} \Theta_{2002} \\ \Theta_{3113} \\ \Theta_{0220} \\ \Theta_{1331} \end{pmatrix}(\tau, \zeta), \qquad (63)$$

$$\begin{pmatrix} \Theta_{2002} \\ \Theta_{3113} \\ \Theta_{0220} \\ \Theta_{1331} \end{pmatrix}(\tau, A\zeta) = \frac{1}{2}\begin{pmatrix} +1 & -1 & +1 & +1 \\ -1 & +1 & +1 & +1 \\ +1 & +1 & +1 & -1 \\ -1 & -1 & +1 & -1 \end{pmatrix}\begin{pmatrix} \Theta_{2200} \\ \Theta_{3311} \\ \Theta_{0022} \\ \Theta_{1133} \end{pmatrix}(\tau, \zeta), \qquad (64)$$

$$\begin{pmatrix} \Theta_{3300} \\ \Theta_{2211} \\ \Theta_{1122} \\ \Theta_{0033} \end{pmatrix}(\tau, A\zeta) = \frac{1}{2}\begin{pmatrix} +1 & -1 & +1 & +1 \\ -1 & +1 & +1 & +1 \\ -1 & -1 & -1 & +1 \\ +1 & +1 & -1 & +1 \end{pmatrix}\begin{pmatrix} \Theta_{3030} \\ \Theta_{2121} \\ \Theta_{1212} \\ \Theta_{0303} \end{pmatrix}(\tau, \zeta), \qquad (65)$$

$$\begin{pmatrix} \Theta_{3030} \\ \Theta_{2121} \\ \Theta_{1212} \\ \Theta_{0303} \end{pmatrix}(\tau, A\zeta) = \frac{1}{2}\begin{pmatrix} +1 & -1 & +1 & +1 \\ -1 & +1 & +1 & +1 \\ -1 & -1 & -1 & +1 \\ +1 & +1 & -1 & +1 \end{pmatrix}\begin{pmatrix} \Theta_{3003} \\ \Theta_{2112} \\ \Theta_{1221} \\ \Theta_{0330} \end{pmatrix}(\tau, \zeta), \qquad (66)$$

$$\begin{pmatrix} \Theta_{3003} \\ \Theta_{2112} \\ \Theta_{1221} \\ \Theta_{0330} \end{pmatrix}(\tau, A\zeta) = \frac{1}{2}\begin{pmatrix} +1 & -1 & +1 & +1 \\ -1 & +1 & +1 & +1 \\ -1 & -1 & -1 & +1 \\ +1 & +1 & -1 & +1 \end{pmatrix}\begin{pmatrix} \Theta_{3300} \\ \Theta_{2211} \\ \Theta_{1122} \\ \Theta_{0033} \end{pmatrix}(\tau, \zeta). \qquad (67)$$

These identities among four-variable theta functions can then be turned into identities among the one-variable theta functions using the relation

$$\Theta_{jklm}(\tau, (\zeta_1, \zeta_2, \zeta_3, \zeta_4)^T) = \theta_j(\tau, \zeta_1)\theta_k(\tau, \zeta_2)\theta_l(\tau, \zeta_3)\theta_m(\tau, \zeta_4), \qquad (68)$$

which is proved by essentially the same argument that gave (23).

Exercises

1. Prove the addition formulae

$$\begin{aligned} \theta_0(\tau, 0)^2\theta_0(\tau, \xi + \eta)\theta_0(\tau, \xi - \eta) &= \theta_3(\tau, \xi)^2\theta_3(\tau, \eta)^2 \qquad (69)\\ &\quad - \theta_2(\tau, \xi)^2\theta_2(\tau, \eta)^2 \\ &= \theta_0(\tau, \xi)^2\theta_0(\tau, \eta)^2 \\ &\quad - \theta_1(\tau, \xi)^2\theta_1(\tau, \eta)^2 \end{aligned}$$

$$\begin{aligned} \theta_0(\tau, 0)\theta_2(\tau, 0)\theta_0(\tau, \xi + \eta)\theta_2(\tau, \xi - \eta) &= \theta_0(\tau, \xi)\theta_2(\tau, \xi)\theta_0(\tau, \eta)\theta_2(\tau, \eta) \\ &\quad + \theta_1(\tau, \xi)\theta_3(\tau, \xi)\theta_1(\tau, \eta)\theta_3(\tau, \eta) \quad (70) \end{aligned}$$

$$\begin{aligned} \theta_0(\tau, 0)\theta_3(\tau, 0)\theta_0(\tau, \xi + \eta)\theta_3(\tau, \xi - \eta) &= \theta_0(\tau, \xi)\theta_3(\tau, \xi)\theta_0(\tau, \eta)\theta_3(\tau, \eta) \\ &\quad + \theta_1(\tau, \xi)\theta_2(\tau, \xi)\theta_1(\tau, \eta)\theta_2(\tau, \eta) \quad (71) \end{aligned}$$

$$\begin{aligned} \theta_3(\tau, 0)\theta_2(\tau, 0)\theta_1(\tau, \xi + \eta)\theta_0(\tau, \xi - \eta) &= \theta_0(\tau, \xi)\theta_1(\tau, \xi)\theta_3(\tau, \eta)\theta_2(\tau, \eta) \\ &\quad + \theta_3(\tau, \xi)\theta_2(\tau, \xi)\theta_0(\tau, \eta)\theta_1(\tau, \eta) \quad (72) \end{aligned}$$

by finding appropriate specializations of $(\zeta_1, \zeta_2, \zeta_3, \zeta_4)$.

2. Prove the formulae

$$\theta_3(\tau, 0)^2\theta_3(\tau, \xi)^2 = \theta_0(\tau, 0)^2\theta_0(\tau, \xi)^2 + \theta_2(\tau, 0)^2\theta_2(\tau, \xi)^2 \qquad (73)$$

$$\theta_3(\tau, 0)^2\theta_0(\tau, \xi)^2 = \theta_0(\tau, 0)^2\theta_3(\tau, \xi)^2 + \theta_2(\tau, 0)^2\theta_1(\tau, \xi)^2 \qquad (74)$$

$$\theta_3(\tau, 0)^2\theta_2(\tau, \xi)^2 = \theta_2(\tau, 0)^2\theta_3(\tau, \xi)^2 - \theta_0(\tau, 0)^2\theta_1(\tau, \xi)^2 \qquad (75)$$

$$\theta_3(\tau, 0)^2\theta_1(\tau, \xi)^2 = \theta_2(\tau, 0)^2\theta_0(\tau, \xi)^2 - \theta_0(\tau, 0)^2\theta_2(\tau, \xi)^2 \qquad (76)$$

by finding appropriate specializations of $(\zeta_1, \zeta_2, \zeta_3, \zeta_4)$.

3. Prove that

$$\prod_{k=1}^{\infty}(1 - q^k) = \sum_{n=-\infty}^{\infty} (-1)^n q^{\frac{3}{2}n^2+\frac{1}{2}n}$$

Hint: Write down sum and product expansions for

$$\theta_0(\frac{3}{2}\tau, \frac{1}{2}\tau).$$

4. Let $p(n)$ be the number of *partitions* of n, that is, the number of ways that n can be written as a sum of positive integers, not counting as distinct those which differ only in the order of the summands. Prove the following formula from the section "Partitio numerorum" of [26],

$$p(n) = p(n-1) + p(n-2) - p(n-5) - p(n-7) + p(n-12) + p(n-15) - \cdots$$

Hint: First show that

$$\prod_{k=1}^{\infty}(1 - q^k) \sum_{n=0}^{\infty} p(n)q^n = 1.$$

Then apply the preceding exercise.

5. In Section 3.2 we saw that

$$\theta_1(\frac{a\tau + b}{c\tau + d}, \frac{\zeta}{c\tau + d}) = (c\tau + d)^{-1} \frac{\frac{\partial\theta_1(\frac{a\tau + b}{c\tau+d}, 0)}{\partial\zeta}}{\frac{\partial\theta_1(\tau, 0)}{\partial\zeta}} e^{\pi i \frac{c\zeta^2}{c\tau+d}} \theta_1(\tau, \zeta).$$

when a, b, c, and d are integers satisfying $ad - bc = 1$. Show that

$$\theta_1(\frac{a\tau + b}{c\tau + d}, \frac{\zeta}{c\tau + d}) = C(c\tau + d)^{1/2}e^{\pi i \frac{c\zeta^2}{c\tau+d}}, \theta_1(\tau, \zeta). \qquad (77)$$

Where C depends only on a, b, c, and d.

6. The matrix A had the property that

$$AL^{(m)} = L^{(m)}$$

for all integers m. How many other matrices have this property?

3.4   Elliptic Functions

We now consider quotients of theta functions in some detail. These are examples
of *elliptic functions*, that is doubly periodic meromorphic functions. They were
discovered independently by Abel [1] and Jacobi [38].

Let[7]

$$\kappa(\tau) = \frac{\theta_2(\tau, 0)^2}{\theta_3(\tau, 0)^2}, \quad \kappa'(\tau) = \frac{\theta_0(\tau, 0)^2}{\theta_3(\tau, 0)^2}, \tag{1}$$

$$K(\tau) = \frac{\pi}{2}\theta_3(\tau, 0)^2, \quad K'(\tau) = \frac{-i\pi\tau}{2}\theta_3(\tau, 0)^2, \tag{2}$$

$$s(u, \tau) = \frac{\theta_3(\tau, 0)}{\theta_2(\tau, 0)}\frac{\theta_1(\tau, \zeta)}{\theta_0(\tau, \zeta)},$$

$$c(u, \tau) = \frac{\theta_0(\tau, 0)}{\theta_2(\tau, 0)}\frac{\theta_2(\tau, \zeta)}{\theta_0(\tau, \zeta)}, \tag{3}$$

$$d(u, \tau) = \frac{\theta_0(\tau, 0)}{\theta_3(\tau, 0)}\frac{\theta_3(\tau, \zeta)}{\theta_0(\tau, \zeta)},$$

where $\zeta = \frac{u}{2K(\tau)}$. Our detailed knowledge of theta functions allows us to say a
great deal about the functions s, c, and d.

THEOREM 31 (BASIC PROPERTIES OF ELLIPTIC FUNCTIONS)
The functions s, c, and d satisfy the algebraic relations

$$s(u, \tau)^2 + c(u, \tau)^2 = 1 \tag{4}$$

and

$$\kappa(\tau)^2 s(u, \tau)^2 + d(u, \tau)^2 = 1, \tag{5}$$

the differential equations

$$\frac{\partial}{\partial u}s(u, \tau) = c(u, \tau)d(u, \tau), \tag{6}$$

$$\frac{\partial}{\partial u}c(u, \tau) = -s(u, \tau)d(u, \tau),$$

$$\frac{\partial}{\partial u}d(u, \tau) = -\kappa(\tau)^2 s(u, \tau)c(u, \tau),$$

---

[7]The notation κ, κ', K, K' is standard. Note that the primes here do not denote differentiation.
The notation s, c, and d is nonstandard, and will soon be replaced by Jacobi's notation.

and the addition formulae

$$s(u+v,\tau) = \frac{s(u,\tau)c(v,\tau)d(v,\tau)+s(v,\tau)c(u,\tau)d(u,\tau)}{1-\kappa(\tau)^2s(u,\tau)^2s(v,\tau)^2}, \tag{7}$$

$$c(u+v,\tau) = \frac{c(u,\tau)c(v,\tau)-s(u,\tau)s(v,\tau)d(u,\tau)d(v,\tau)}{1-\kappa(\tau)^2s(u,\tau)^2s(v,\tau)^2},$$

$$d(u+v,\tau) = \frac{d(u,\tau)d(v,\tau)-\kappa(\tau)^2s(u,\tau)s(v,\tau)c(u,\tau)c(v,\tau)}{1-\kappa(\tau)^2s(u,\tau)^2s(v,\tau)^2}.$$

All functions are doubly periodic meromorphic functions of u with periods $4K(\tau)$ and $4iK'(\tau)$. All have simple poles at u of the form $miK'(\tau)+nK(\tau)$ for m odd and n even. They have no other poles. They have simple zeroes at the points $u=miK'(\tau)+nK(\tau)$ where m, n are both even for s, both odd for d, and where m is even and n odd for c. Each function has an additional period. These are $2iK'(\tau)$ for s, $2iK'(\tau)+2K(\tau)$ for c, and $2K(\tau)$ for d.

Proof.   All of these follow immediately from Theorem 28 or from the theta relations of the preceding section.

We now find an important integral representation for *s*.

THEOREM 32 (AN ELLIPTIC INTEGRAL)
For each τ there is a $\rho(\tau) > 0$ and a closed contour C such that

$$s(u,\tau) = \frac{1}{2\pi i}\int_C \frac{zy(z)\,dz}{\int_0^z y(x)\,dx - u} \tag{8}$$

for all $|u| < r(\tau)$, where

$$y(x) = (1-x^2)^{-1/2}(1-\kappa(\tau)^2x^2)^{-1/2} \tag{9}$$

Proof.   Since $s(0,\tau)=0$ we can find a $\rho > 0$ such that

$$|s(u,\tau)| \leq \frac{1}{2}\min(1,|\kappa(\tau)|^{-1}) \tag{10}$$

for all $|u| \leq \rho$. For such *u* equations (4) and (5) give

$$c(u,\tau) = \sqrt{1-s(u,\tau)^2}, \tag{11}$$

$$d(u,\tau) = \sqrt{1-\kappa(\tau)^2s(u,\tau)^2}$$

with the principal values of the square root. From equation (6) we see that

$$\frac{\partial}{\partial u}s(u,\tau) = (1-s(u,\tau)^2)^{1/2}(1-\kappa(\tau)^2s(u,\tau)^2)^{1/2}. \tag{12}$$

In particular $\frac{\partial}{\partial u}s(0, \tau) = 1 \neq 0$. The inverse function theorem therefore shows that there is an R with $0 < R < \rho$ such that s has a well-defined analytic inverse f on the disc of radius R,

$$f(s(u, \tau)) = u. \tag{13}$$

Let $r > 0$ be smaller than

$$\max_{|u|<R} |s(u, \tau)|$$

and let C be the counterclockwise oriented circle of radius R centered at 0.
  By (12) and the chain rule

$$\frac{df}{dx} = (1 - x^2)^{-1/2}(1 - \kappa(\tau)^2 x^2)^{-1/2} = y(x). \tag{14}$$

Since $f(0) = 0$ the fundamental theorem of calculus shows that

$$f(z) = \int_0^z y(x)\, dx. \tag{15}$$

Integrals of the form (15) are called elliptic integrals of the first kind. Historically the study of elliptic functions began with these integrals.
  What we want is not f but its inverse. We can find this using the calculus of residues,

$$f^{-1}(u) = \frac{1}{2\pi i} \int_C \frac{z f'(z)\, dz}{f(z) - u} \tag{16}$$

for all u such that $\mathrm{Ind}(C, f^{-1}(u)) = 1$. Here we choose any C contained in the disc of radius ρ with $\mathrm{Ind}(C, 0) = 1$. We call that connected component of the complement of C which contains 0, U. For our f this gives exactly equation (8) and the proof of the theorem is complete.
  An interesting feature of equation (8) is that $s(u, \tau)$ does not depend directly on τ, but only on $\kappa(\tau)^2$. In other words if

$$\kappa(\tau)^2 = \kappa(\tau')^2 \tag{17}$$

for some τ, τ' then there is a neighborhood U of 0 in which

$$s(u, \tau) = s(u, \tau') \tag{18}$$

and hence, using equations (11),

$$\begin{aligned} c(u, \tau) &= c(u, \tau'), \\ d(u, \tau) &= d(u, \tau'). \end{aligned} \tag{19}$$

Initially we know only that equations (18) and (19) hold for $|u| < r$. We can extend them to all of $\mathbf{C}$ by rigidity. A more direct proof can be given using the addition formulae (7). If (18) and (19) hold for all u with $|u| < r$ then they must hold for all $u + v$ with $|u| < r$ and $|v| < r$. It follows that they hold for all u with $|u| < 2r$. Inductively (18) and (19) hold $|u| < 4r$, $|u| < 8r$, and so forth. Since (17) implies (18) and (19), it makes sense to define functions sin am, cos am, and Δ am[8] by the equations

$$s(u, \tau) = \sin \text{am}(u, \kappa(\tau)), \qquad (20)$$

$$c(u, \tau) = \cos \text{am}(u, \kappa(\tau)),$$

$$d(u, \tau) = \Delta \, \text{am}(u, \kappa(\tau)).$$

We could, of course, rephrase Theorem 31 in terms of sin am, cos am, and Δ am, but it hardly seems worth the effort.

In contrast to s, c, and d, the functions $K(\tau)$ and $iK'(\tau)$ depend directly on τ, not merely on $\kappa^2(\tau)$. The group of translations they generate, however, depends only on $\kappa^2(\tau)$. In principle one could check this using the transformation formulae for the theta functions, but this misses the point. The group of translations generated by $4K(\tau)$ and $4iK'(\tau)$ is simply the largest group of translations which leave s, c, and d invariant. This follows immediately from Theorem 31. Since the functions s, c, and d depend on τ only through $\kappa^2(\tau)$, it follows that the group generated by $4K(\tau)$ and $4iK'(\tau)$ depends only on $\kappa^2(\tau)$. If τ and τ' are such that

$$\kappa^2(\tau) = \kappa^2(\tau') \qquad (21)$$

then $4K(\tau')$ and $4iK'(\tau')$ must be related to $4K(\tau)$ and $4iK'(\tau)$ by an equation of the form

$$\begin{pmatrix} 4iK' \\ 4K \end{pmatrix} (\tau') = \begin{pmatrix} \alpha & \beta \\ \gamma & \delta \end{pmatrix} \begin{pmatrix} 4iK' \\ 4K \end{pmatrix} (\tau) \qquad (22)$$

with α, β, γ, δ integers with $\alpha\delta - \beta\gamma = 1$. From the definition of K and K' however we see that

$$\tau = \frac{4iK'(\tau)}{4K(\tau)} \qquad \tau' = \frac{4iK'(\tau')}{4K(\tau')}, \qquad (23)$$

so

$$\tau' = \frac{\alpha\tau + \beta}{\gamma\tau + \delta}. \qquad (24)$$

---

[8]These somewhat unwieldy names are used throughout the nineteenth-century literature. They derive from Jacobi's observation that, for real u, there is a well-defined function am(u, κ) such that sin am(u, κ), cos am(u, κ), and Δ am(u, κ) are, respectively, the sine, cosine, and derivative of am(u, κ). Most modern books use the notation sn, cn, and dn, introduced by Gudermann.

This argument can be refined slightly. Not only does the group of translations preserving all three functions, s, c, and d, depend only on $\kappa^2(\tau)$, the same is true of the group preserving any one of them. The translations preserving s, for example, are generated by $4K(\tau)$ and $2iK'(\tau)$. It follows then that, if τ and τ' are such that (21) holds, then

$$\begin{pmatrix} 2iK' \\ 4K \end{pmatrix}(\tau') = \begin{pmatrix} \alpha_s & \beta_s \\ \gamma_s & \delta_s \end{pmatrix} \begin{pmatrix} 2iK' \\ 4K \end{pmatrix}(\tau) \tag{25}$$

with α_s, β_s, γ_s, δ_s integers and $\alpha_s\delta_s - \beta_s\gamma_s = 1$. Using (23) again we find

$$\frac{1}{2}\tau' = \frac{\alpha_s\frac{1}{2}\tau + \beta_s}{\gamma_s\frac{1}{2}\tau + \delta_s} \tag{26}$$

or

$$\tau' = \frac{\alpha_s\tau + 2\beta_s}{\frac{1}{2}\gamma_s\tau + \delta_s}. \tag{27}$$

This is compatible with (24) if and only if β is even. A similar argument, using the function d instead of s, shows that γ must be even as well. Using c gives no further information. Of course, if β and γ are even and $\alpha\delta - \beta\gamma = 1$ then α and δ must be odd. In other words the matrix $\begin{pmatrix} \alpha & \beta \\ \gamma & \delta \end{pmatrix}$ is congruent, modulo 2, to the identity matrix. Using the transformation formulae for the theta functions, one can prove the converse: If τ and τ' are related by (24) where $\begin{pmatrix} \alpha & \beta \\ \gamma & \delta \end{pmatrix}$ is congruent, modulo 2, to the identity matrix then $\kappa^2(\tau) = \kappa^2(\tau')$.

What we have just shown is that κ^2 is a modular function [9] of level 2. This is of

---

[9] A *modular function* of *level n* is an analytic function $g(\tau)$ defined on the upper half plane $\text{Im}\,\tau > 0$ satisfying the growth condition

$$|g(\tau)| \leq Ce^{k\,\text{Im}\,\tau}$$

with some positive constants C and k and the functional equations

$$g(\frac{a\tau + b}{c\tau + d}) = g(\tau)$$

for any integers a, b, c, and d with

$$ad - bc = 1$$

and

$$\begin{pmatrix} a & b \\ c & d \end{pmatrix} \equiv \begin{pmatrix} 1 & 0 \\ 0 & 1 \end{pmatrix} \quad \text{mod. } n$$

independent interest, but the reason for proving it now is to allow an unambiguous statement of the following theorem.

Exercises

1. Show that

$$\sin \text{am}(u + iK', \kappa) = \frac{1}{\kappa \sin \text{am}(u, \kappa)},$$

$$\cos \text{am}(u + iK', \kappa) = \frac{-i \Delta \text{am}(u, \kappa)}{\kappa \sin \text{am}(u, \kappa)},$$

and

$$\Delta \text{am}(u + iK', \kappa) = \frac{-i \cos \text{am}(u, \kappa)}{\sin \text{am}(u, \kappa)},$$

2. Show that

$$\sin \text{am}(iu, \kappa) = i \frac{\sin \text{am}(u, \kappa')}{\cos \text{am}(u, \kappa')},$$

$$\cos \text{am}(iu, \kappa) = \frac{1}{\cos \text{am}(u, \kappa')},$$

$$\Delta \text{am}(iu, \kappa) = \frac{\Delta \text{am}(u, \kappa')}{\cos \text{am}(u, \kappa')}.$$

3. Show that

$$\sin \text{am}(\kappa u, \kappa^{-1}) = \kappa \sin \text{am}(u, \kappa),$$

$$\cos \text{am}(\kappa u, \kappa^{-1}) = \Delta \text{am}(u, \kappa),$$

and

$$\Delta \text{am}(\kappa u, \kappa^{-1}) = \cos \text{am}(u, \kappa).$$

3.5 Complex Multiplication

This section contains a very brief introduction to the theory of complex multiplication, which relates the analysis of elliptic functions to the arithmetic of imaginary quadratic fields. Here we treat only the analytic aspects of the subject. Readers interested in pursuing the subject further should consult Kronecker's paper [43] and Weber's book [72].

We have already considered how our functions transform when τ is replaced by $\tau' = \frac{a\tau+b}{c\tau+d}$ where a, b, c, and d and $ad - bc = 1$. We now replace this last assumption by

$$ad - bc = L, \tag{1}$$

where L is a positive integer. The assumption that L is positive is needed to ensure that $\operatorname{Im} \tau' > 0$.

Let

$$\mu(\tau) = (c\tau + d)\frac{K(\tau)}{K(\tau')}. \tag{2}$$

μ is an analytic function on the upper half plane $\operatorname{Im} \tau > 0$ depending on a, b, c, and d. It has no zeroes because $\theta_3(\tau, 0)$ is never zero. From equation (2) of Section 3.4 we see that

$$\frac{iK'(\tau')}{K(\tau')} = \tau' = \frac{a\tau + b}{c\tau + d} = \frac{ai K'(\tau) + bK(\tau)}{ci K'(\tau) + dK(\tau)} \tag{3}$$

and hence

$$\begin{pmatrix} a & b \\ c & d \end{pmatrix} \begin{pmatrix} iK' \\ K \end{pmatrix}(\tau) = \mu(\tau) \begin{pmatrix} iK' \\ K \end{pmatrix}(\tau'). \tag{4}$$

We will adopt the following notational conventions. κ, K, K', and μ without arguments will always be understood to be evaluated at τ. λ, Λ, Λ' will be used for $\kappa(\tau')$, $K(\tau')$, and $K'(\tau')$ respectively. Thus we will write the previous equation as

$$\begin{pmatrix} a & b \\ c & d \end{pmatrix} \begin{pmatrix} iK' \\ K \end{pmatrix} = \mu \begin{pmatrix} i\Lambda' \\ \Lambda \end{pmatrix}. \tag{5}$$

We will need the following simple fact from linear algebra, which we will not prove.

THEOREM 33

Let a, b, c, and d be integers satisfying (1) for some positive L. There are integers $r_1, \ldots, r_L$ and $s_1, \ldots, s_L$ such that every $(m, n) \in \mathbb{Z}^2$ can be written uniquely as

$$(m \quad n) = -(r_j \quad s_j) + (m' \quad n') \begin{pmatrix} a & b \\ c & d \end{pmatrix} \tag{6}$$

with $1 \le j \le L$ and $(m' \quad n') \in \mathbb{Z}^2$. We may arrange that $(r_1, s_1) = (0 \quad 0)$.

The main result of this section is the following theorem.

THEOREM 34 (JACOBI'S FORMULA FOR COMPLEX MULTIPLICATION)

With notation as above, suppose that a and d are odd and that b is even. Then

$$\operatorname{sin am}(\mu u, \kappa) = \mu \operatorname{sin am}(u, \kappa) \prod_{j=2}^{N} \frac{\operatorname{sin am}(u + \frac{2r_j i K' + 2s_j K}{\mu}, \lambda)}{\operatorname{sin am}(\frac{2r_j i K' + 2s_j K}{\mu}, \lambda)}. \tag{7}$$

This theorem is particularly interesting when $\kappa = \lambda$, as we will see in Section 3.7.

Proof.   We begin by computing some logarithmic derivatives. Equation (56) of Section 3.2 says that[10]

$$\frac{\partial}{\partial \zeta} \log \theta_1(\tau, \zeta) = \lim_{M \to \infty} \lim_{N \to \infty} \sum_{\substack{-M \le m \le M \\ -N \le n \le N}} (\zeta - m\tau - n)^{-1}. \tag{8}$$

To improve the readability of the following formulae, we will assume that all otherwise unspecified sums and limits are as in equation (8) The translation relations, equation (3) of Section 3.1, then show that

$$\frac{\partial}{\partial \zeta} \log \theta_0(\tau, \zeta) = \lim \sum (\zeta - (m - \frac{1}{2})\tau - n)^{-1} + \pi i \tag{9}$$

and hence

$$\frac{\partial}{\partial \zeta} \log \frac{\theta_3(\tau, 0) \, \theta_1(\tau, \zeta)}{\theta_2(\tau, 0) \, \theta_0(\tau, \zeta)} = \lim \sum [(\zeta - m\tau - n)^{-1} \tag{10}$$

$$- (\zeta - (m - \frac{1}{2})\tau - n)^{-1}]$$

$$- \pi i.$$

---

[10]Recall that the logarithmic derivative of θ_1 is well-defined even though $\log \theta_1$ is not.

In terms of elliptic functions equation (10) says that

$$\frac{\partial}{\partial u} \log \sin \mathrm{am}(u, \kappa) = \lim \sum [(u - 2mi K' - 2nK)^{-1}$$

$$- (u - (2m - 1)i K' - 2nK)^{-1}]$$

$$- \frac{\pi i}{2K},$$

where the limit and sum are again as in (8). Taking two derivatives we find

$$\frac{\partial^3}{\partial u^3} \log \sin \mathrm{am}(u, \kappa) = 2 \lim \sum [(u - 2mi K' - 2nK)^{-3} \tag{12}$$

$$- (u - (2m - 1)i K' - 2nK)^{-3}],$$

or, since the sums are absolutely convergent,

$$\frac{\partial^3}{\partial u^3} \log \sin \mathrm{am}(u, \kappa) = 2 \sum_{(m,n) \in \mathbf{Z}^2} (u - 2mi K' - 2nK)^{-3}$$

$$- 2 \sum_{(m,n) \in \mathbf{Z}^2} (u - (2m - 1)i K' - 2nK)^{-3}, \tag{13}$$

where the order in which the sum is taken is immaterial. It follows that

$$\frac{\partial^3}{\partial u^3} \log \sin \mathrm{am}(\mu u, \kappa) = 2 \sum_{(m,n) \in \mathbf{Z}^2} (u - 2mi \mu^{-1} K' - 2n\mu^{-1} K)^{-3} \tag{14}$$

$$- 2 \sum_{(m,n) \in \mathbf{Z}^2} (u - (2m - 1)\mu^{-1}i K' - 2n\mu^{-1} K)^{-3}.$$

We evaluate the sums on the right with the aid of Theorem 33.

$$\mu^{-1} \begin{pmatrix} m & n \end{pmatrix} \begin{pmatrix} i K' \\ K \end{pmatrix} = \begin{pmatrix} m & n \end{pmatrix} \begin{pmatrix} a & b \\ c & d \end{pmatrix}^{-1} \begin{pmatrix} i \Lambda' \\ \Lambda \end{pmatrix} \tag{15}$$

$$= -(r_j, s_j) \begin{pmatrix} a & b \\ c & d \end{pmatrix}^{-1} \begin{pmatrix} i \Lambda' \\ \Lambda \end{pmatrix} + \begin{pmatrix} m' & n' \end{pmatrix} \begin{pmatrix} i \Lambda' \\ \Lambda \end{pmatrix}$$

$$= -\mu^{-1} (r_j, s_j) \begin{pmatrix} i K' \\ K \end{pmatrix} + \begin{pmatrix} m' & n' \end{pmatrix} \begin{pmatrix} i \Lambda' \\ \Lambda \end{pmatrix}$$

and thus

$$\mu^{-1} \begin{pmatrix} 2m & 2n \end{pmatrix} \begin{pmatrix} i K' \\ K \end{pmatrix} = -(2r_j, 2s_j) \begin{pmatrix} \mu^{-1}i K' \\ \mu^{-1} K \end{pmatrix} + \begin{pmatrix} 2m' & 2n' \end{pmatrix} \begin{pmatrix} i \Lambda' \\ \Lambda \end{pmatrix}. \tag{16}$$

We can therefore replace the sum

$$\sum_{(m,n)\in\mathbb{Z}^2} (u - 2mi\mu^{-1}K' - 2n\mu^{-1}K)^{-3}$$

in equation (14) by

$$\sum_{j=1}^{L} \sum_{(m',n')\in\mathbb{Z}^2} \left(u + \frac{2r_j K' + 2s_j K}{\mu} - 2m'i\Lambda' - 2n'\Lambda\right)^{-3}.$$

The argument for the other term in (14) is similar, but more complicated. We define

$$m'' = m + \frac{a-1}{2}, \quad n'' = n + \frac{b}{2}. \tag{17}$$

Since a is odd and b even $(m'', n'') \in \mathbb{Z}^2$ and we may apply Theorem 34 to show that

$$\begin{pmatrix} m'' & n'' \end{pmatrix} = -\begin{pmatrix} r_j & s_j \end{pmatrix} + \begin{pmatrix} m' & n' \end{pmatrix} \begin{pmatrix} a & b \\ c & d \end{pmatrix} \tag{18}$$

with $1 \le j \le L$ and $\begin{pmatrix} m' & n' \end{pmatrix} \in \mathbb{Z}^2$, and hence

$$\begin{pmatrix} 2m-1 & 2n \end{pmatrix} = -\begin{pmatrix} r_j & s_j \end{pmatrix} + \begin{pmatrix} 2-1m' & n' \end{pmatrix} \begin{pmatrix} a & b \\ c & d \end{pmatrix}. \tag{19}$$

It follows that

$$\mu^{-1} \begin{pmatrix} 2m-1 & 2n \end{pmatrix} \begin{pmatrix} iK' \\ K \end{pmatrix} = -\begin{pmatrix} 2r_j, 2s_j \end{pmatrix} \begin{pmatrix} \mu^{-1}iK' \\ \mu^{-1}K \end{pmatrix} \tag{20}$$

$$+ \begin{pmatrix} 2m'-1 & 2n' \end{pmatrix} \begin{pmatrix} i\Lambda' \\ \Lambda \end{pmatrix}.$$

We can therefore replace the sum

$$\sum_{(m,n)\in\mathbb{Z}^2} (u - (2m-1)i\mu^{-1}K' - 2n\mu^{-1}K)^{-3}$$

in equation (14) by

$$\sum_{j=1}^{L} \sum_{(m',n')\in\mathbb{Z}^2} (u + \frac{2r_j K' + 2s_j K}{\mu} - (2m'-1)i\Lambda' - 2n'\Lambda)^{-3}.$$

Now we sum over m' and n', obtaining

$$\frac{\partial^3}{\partial u^3} \log \sin \mathrm{am}(u, \kappa) = \sum_{j=1}^{L} \frac{\partial^3}{\partial u^3} \log \sin \mathrm{am}(u + \frac{2r_j i K' + 2s_j K}{\mu}, \lambda). \quad (21)$$

Let $f(u)$ denote the quotient of the right hand side of (7) by the left hand side. Equation (21) shows that

$$\frac{\partial^3}{\partial u^3} \log f(u) = 0 \quad (22)$$

and hence

$$f(u) = C e^{\alpha u + \beta u^2}. \quad (23)$$

for some constants C, α, and β. But f is periodic with periods $4\mu^{-1}K$ and $4\mu^{-1}iK'$, so

$$f(u + p) = f(u), \quad (24)$$

whenever $u \in \mathbf{C}$ and $p = 4m\mu^{-1}iK' + 4n\mu^{-1}K$. From this equation it follows that

$$e^{\alpha p + \beta p^2 + 2\beta pu} = 1. \quad (25)$$

The only way this can hold for all u is if $\beta = 0$. But then

$$e^{\alpha p} = 1 \quad (26)$$

for all $p = 4m\mu^{-1}iK' + 4n\mu^{-1}K$. Letting $(m, n) = (0, 1)$ we find that

$$\alpha = \frac{\mu l \pi i}{2K} \quad (27)$$

for some integer l. Taking $(m, n) = (1, 0)$ we find

$$e^{\frac{-2\pi l K'}{K}} = e^{2\pi i l \tau} = 1, \quad (28)$$

which is impossible unless $l = 0$ and hence $\alpha = 0$. L'Hôpital's rule shows that $f(u) = 1$ so $C = 1$, and the proof of Theorem 34 is complete.

3.6   Quadratic Reciprocity

This section essentially follows Eisenstein's 1846 paper [20], with some minor improvements from Kronecker's 1886 paper [43]. We derive quadratic reciprocity from the properties of the trigonometric functions. In Section 3.7 we will derive biquadratic reciprocity from the corresponding properties of the lemniscatic functions. These latter are just the elliptic functions for $\kappa = i$.

We begin by defining the quadratic character. Let $p \in \mathbf{Z}$ be positive and odd. Let $F = \mathbf{Z}/p\mathbf{Z}$. We are not assuming p to be prime, so F need not be a field. Let $U = \{1, -1\}$ be the group of units of $\mathbf{Z}$. U acts on the set $F - \{0\}$ without fixed points so $F - \{0\}$ splits into $\frac{p-1}{2}$ U-orbits, each of cardinality 2. We choose a set $R \subset \mathbf{Z}$ of representatives for these orbits. The simplest such choice is $R = \{1, 2, \ldots, \frac{p-1}{2}\}$. If q is relatively prime to p and $r \in R$ then

$$qr \equiv \sigma_p(q, r)r' \quad \text{mod. } p \tag{1}$$

for a unique $\sigma_p(q, r)$ in U and r' in R. We define the *quadratic character modulo* p, denoted by the Legendre symbol $\left(\frac{q}{p}\right)$, by

$$\left(\frac{q}{p}\right) = \prod_{r \in R} \sigma_p(q, r). \tag{2}$$

It is not hard to show that $\left(\frac{q}{p}\right)$ is independent of the way in which the set R is chosen. It is clear from the definitions that $\left(\frac{q}{p}\right) \in U$, and that $\left(\frac{q}{p}\right)$ depends only on the class of q modulo p.

The definition (2) is not a practical way to calculate $\left(\frac{q}{p}\right)$ unless p or q is very small. Two special cases are of particular interest. If $q = -1$ then $\sigma_p(q, r) = -1$ for all r, so $\left(\frac{q}{p}\right)$ is -1 to the cardinality of R. In other words,

$$\left(\frac{-1}{p}\right) = (-1)^{\frac{p-1}{2}}. \tag{3}$$

If $q = 2$ then $\sigma_p(q, r) = 1$, if $1 \leq 2r \leq \frac{p-1}{2}$, and $\sigma_p(q, r) = -1$, if $\frac{p+1}{2} \leq 2r \leq p - 1$. The number of -1's is easily seen to be even, if $p \equiv \pm 1$ mod. 8, and odd, if $p \equiv \pm 3$ mod. 8. It follows that

$$\left(\frac{2}{p}\right) = \begin{cases} 1 & \text{if } p \equiv \pm 1 \quad \text{mod. 8} \\ -1 & \text{if } p \equiv \pm 3 \quad \text{mod. 8} \end{cases}. \tag{4}$$

If $q = q_1 q_2$ then

$$q_2 r \equiv \sigma_p(q_2, r)r' \quad \text{mod. } p \tag{5}$$

and

$$q_1 r' \equiv \sigma_p(q_1, r')r'' \quad \text{mod. } p \tag{6}$$

so

$$qr \equiv \sigma_p(q_2, r)\sigma_p(q_1, r')r'' \quad \text{mod. } p \tag{7}$$

and

$$\sigma_p(q, r) \equiv \sigma_p(q_2, r)\sigma_p(q_1, r') \quad \text{mod. } p. \tag{8}$$

Both sides are ± 1, so the congruence can hold only if

$$\sigma_p(q, r) = \sigma_p(q_2, r)\sigma_p(q_1, r'). \tag{9}$$

As r ranges over R so does r'. The product of (8) taken over all $r \in R$ is then

$$\left(\frac{q}{p}\right) = \left(\frac{q_1}{p}\right)\left(\frac{q_2}{p}\right). \tag{10}$$

The Legendre symbol is multiplicative, takes only the values ± 1, and depends only on the residue of q modulo p, so we are justified in calling it a quadratic character modulo p.

If p is prime then F is a field, and $F - \{0\}$ a cyclic group of order $p - 1$. Taking the product of (1) over all $r \in R$ we find

$$q^{\frac{p-1}{2}} \prod_{r \in R} r \equiv \left(\frac{q}{p}\right) \prod_{r \in R} r' \quad \text{mod. } p. \tag{11}$$

As r ranges over R so does r', so

$$\prod_{r \in R} r \equiv \prod_{r \in R} r \quad \text{mod. } p. \tag{12}$$

This product is invertible in F, so

$$q^{\frac{p-1}{2}} \equiv \left(\frac{q}{p}\right) \quad \text{mod. } p. \tag{13}$$

Since $F - \{0\}$ is cyclic of order $p - 1$, we see that $\left(\frac{q}{p}\right) = 1$, if q is congruent to a square modulo p, and $\left(\frac{q}{p}\right) = -1$ otherwise.

We must stress that the congruence (13) was proved only under the hypothesis that p is prime. Without this hypothesis it is generally false. Whether or not p is prime, both sides of (13) are easy to compute. This gives a useful primality test. If, for some q, the congruence (13) is violated we know that p is composite, even though we have not produced a nontrivial factor. Converses of various sorts exist, but these are a subject for more specialized texts.

To compute the left hand side we write $\frac{p-1}{2}$ in binary,

$$\frac{p-1}{2} = \sum_{j=0}^{n} c_j 2^j \tag{14}$$

with $c_j \in \{0, 1\}$. Then

$$q^{\frac{p-1}{2}} = \prod_{c_j=1} q^{2^j}. \tag{15}$$

The q^{2^j} are computed inductively using the recurrence

$$q^{2^j} = (q^{2^{j-1}})^2, \tag{16}$$

and all arithmetic is performed modulo p. If one uses the multiplication algorithm taught in elementary schools, the time required is proportional to

$$\log^3 p \log \log p.$$

If one uses a better algorithm, for example the fast Fourier transform, then the time is proportional to

$$\log^2 p \log^2 \log p.$$

The right hand side if (13) is calculated by a variant of the Euclidean algorithm. An essential tool is the following theorem of Gauss.

THEOREM 35 (QUADRATIC RECIPROCITY)
If p and q are relatively prime, positive, odd integers then

$$\left(\frac{q}{p}\right) = (-1)^{\frac{p-1}{2}\frac{q-1}{2}} \left(\frac{p}{q}\right). \tag{17}$$

An example of the usefulness of Theorem 35 is the following calculation. We take $q = 5$, $p = 561$. Quadratic reciprocity shows that

$$\left(\frac{5}{561}\right) = \left(\frac{561}{5}\right) = \left(\frac{1}{5}\right) = 1. \tag{18}$$

On the other hand

$$5^{\frac{561-1}{2}} \equiv -1 \mod. 561, \tag{19}$$

so 561 must be composite. We choose this example not because 561 is hard to factor, indeed $561 = 3 \times 11 \times 17$, but because it is a *Carmichael number*. These are discussed in more detail in the exercises.

Proof. The congruence (1) is simply a concise way to say that there exists an integer a such that

$$qr = \sigma_p(q,r)r' + ap. \tag{20}$$

The sine function is 2π periodic, so

$$\sin \frac{qr}{p} 2\pi = \sin \sigma_p(q,r) \frac{r'}{p} 2\pi. \tag{21}$$

Since $\sigma_p(q,r) = \pm 1$ and sine is odd

$$\sin \frac{qr}{p} 2\pi = \sigma_p(q,r) \sin \frac{r'}{p} 2\pi. \tag{22}$$

We take the product over all $r \in R$,

$$\prod_{r \in R} \sin \frac{qr}{p} 2\pi = \left(\frac{q}{p}\right) \prod_{r \in R} \sin \frac{r'}{p} 2\pi \tag{23}$$

and remember that as r ranges over R so does r',

$$\prod_{r \in R} \sin \frac{qr}{p} 2\pi = \left(\frac{q}{p}\right) \prod_{r \in R} \sin \frac{r}{p} 2\pi. \tag{24}$$

This gives a trigonometric formula for the quadratic character,

$$\left(\frac{q}{p}\right) = \prod_{r \in R} \frac{\sin \frac{qr}{p} 2\pi}{\sin \frac{r}{p} 2\pi}. \tag{25}$$

This is not yet the trigonometric formula we want, but it is a start.

An easy consequence of the product representation for the sine is the relation

$$\sin q\theta = q \sin \theta \prod_s \frac{\sin(\theta + \frac{s}{q}2\pi)}{\sin \frac{s}{q} 2\pi}, \tag{26}$$

where the product is taken over a set of representatives for $\mathbf{Z}/q\mathbf{Z}$ with zero omitted. In analogy with our earlier definition of R, we define S to be a set of representatives for the U orbits of $\mathbf{Z}/q\mathbf{Z} - \{0\}$. We could, for example, take $S = \{1, \ldots, \frac{q-1}{2}\}$. Using equation (26) and the fact that sine is odd,

$$\sin q\theta = (-1)^{\frac{q-1}{2}} q \sin\theta \prod_{s \in S} \frac{\sin(\theta + \frac{s}{q}2\pi)\sin(\theta - \frac{s}{q}2\pi)}{\sin^2 \frac{s}{q}2\pi}. \tag{27}$$

To simplify the numerator in (27) we use the addition formula

$$\sin(\theta + \phi) = \sin\theta\cos\phi + \sin\phi\cos\theta, \tag{28}$$

which implies

$$\sin(\theta + \phi)\sin(\theta - \phi) = \sin^2\theta - \sin^2\phi, \tag{29}$$

to replace (27) by

$$\sin q\theta = (-1)^{\frac{q-1}{2}} q \sin\theta \prod_{s \in S} \frac{\sin^2\theta - \sin^2 \frac{s}{q}2\pi}{\sin^2 \frac{s}{q}2\pi}. \tag{30}$$

The product has $\frac{p-1}{2}$ factors, so multiplying each by -1 gives a factor which exactly cancels the leading $(-1)^{\frac{q-1}{2}}$,

$$\sin q\theta = q \sin\theta \prod_{s \in S} \frac{\sin^2 \frac{s}{q}2\pi - \sin^2\theta}{\sin^2 \frac{s}{q}2\pi}. \tag{31}$$

We need to evaluate the product of the denominators in (31). To this end we set $\theta = \frac{\pi}{2}$ in (30). The translation relation

$$\sin(\theta + \pi) = -\sin\theta \tag{32}$$

gives $\sin q\frac{\pi}{2} = (-1)^{\frac{q-1}{2}}$. Dividing both sides of (30) by $(-1)^{\frac{q-1}{2}}$, we find

$$1 = q \prod_{s \in S} \frac{1 - \sin^2 \frac{s}{q}2\pi}{\sin^2 \frac{s}{q}2\pi} \tag{33}$$

$$= q \prod_{s \in S} \frac{\cos^2 \frac{s}{q}2\pi}{\sin^2 \frac{s}{q}2\pi}.$$

Multiplying both sides of (33) by $\prod_{s \in S} \sin^4 \frac{s}{q}2\pi$ we find

$$\left(\prod_{s \in S} \sin^2 \frac{s}{q}2\pi\right)^2 = q \prod_{s \in S} (\sin\frac{s}{q}2\pi \cos\frac{s}{q}2\pi)^2 \tag{34}$$

or, using the double angle formula,

$$\left(\prod_{s\in S} \sin^2 \frac{s}{q} 2\pi\right)^2 = 2^{1-q} q \prod_{s\in S} \sin^2 \frac{2s}{q} 2\pi. \tag{35}$$

For any $s \in S$ we have

$$2s \equiv \sigma_q(2, s)s' \quad \text{mod. } q \tag{36}$$

with $\sigma_q(2, s) = \pm 1$ and $s' \in S$. Thus

$$\sin^2 \frac{2s}{q} 2\pi = \sin^2 \frac{2s'}{q} 2\pi. \tag{37}$$

As s ranges over S so does s', so

$$\prod_{s\in S} \sin^2 \frac{2s}{q} 2\pi = \prod_{s\in S} \sin^2 \frac{2s'}{q} 2\pi. \tag{38}$$

Applying this to equation (35) we find

$$\prod_{s\in S} \sin^2 \frac{s}{q} 2\pi = 2^{1-q} q. \tag{39}$$

This allows us to rewrite (31) as

$$\sin q\theta = 2^{q-1} \sin \theta \prod_{s\in S} \left(\sin^2 \frac{s}{q} 2\pi - \sin^2 \theta\right). \tag{40}$$

Taking $\theta = \frac{r}{p} 2\pi$ in (40), and then taking the product over all $r \in R$, we find

$$\left(\frac{q}{p}\right) = 2^{\frac{1}{2}(p-1)(q-1)} \prod_{r\in R, s\in S} \left(\sin^2 \frac{s}{q} 2\pi - \sin^2 \frac{r}{p} 2\pi\right). \tag{41}$$

The whole point of this long series of trigonometric identities was to produce equation (41). Reversing the roles of p and q we find

$$\left(\frac{p}{q}\right) = 2^{\frac{1}{2}(p-1)(q-1)} \prod_{r\in R, s\in S} \left(\sin^2 \frac{r}{p} 2\pi - \sin^2 \frac{s}{q} 2\pi\right). \tag{42}$$

Dividing (41) by (42) gives

$$\frac{\left(\frac{q}{p}\right)}{\left(\frac{p}{q}\right)} = \prod_{r\in R, s\in S} (-1). \tag{43}$$

There are $\frac{p-1}{2}\frac{q-1}{2}$ factors in the product, so this just gives equation (17) and the proof of the theorem is complete.

Exercises

1. Show that if p is prime then

$$a^{p-1} \equiv 1 \quad \text{mod. } p$$

for any integer a relatively prime to p. Thus 21 can not be prime because

$$2^{20} \equiv 4 \quad \text{mod. } 21.$$

2. A *Carmichael number* is a composite number n such that

$$a^{n-1} \equiv 1 \quad \text{mod. } n$$

for every integer a relatively prime to n. These numbers cannot therefore be proved composite by the method of the preceding exercise. Show that 561 is a Carmichael number.

3. Another Carmichael number is 1729. Show that it is composite by showing that

$$11^{\frac{1729-1}{2}} \equiv 1 \quad \text{mod. } 1729,$$

while

$$\left(\frac{11}{1729}\right) = -1.$$

3.7 Biquadratic Reciprocity

There is a similar theory with the *Gaussian integers*, $\mathbf{G} = \mathbf{Z}[i]$, in place of $\mathbf{Z}$, biquadratic residues in place of quadratic, and the lemniscatic functions

$$\begin{aligned}
\operatorname{sin lemn} u &= \operatorname{sin am}(u, i), \\
\operatorname{cos lemn} u &= \operatorname{cos am}(u, i), \\
\Delta \operatorname{lemn} u &= \Delta \operatorname{am}(u, i)
\end{aligned} \tag{1}$$

in place of the trigonometric functions. To prevent subsequent calculations from overflowing into the margins, we will call these functions sl, cl, and dl.

We begin by defining the biquadratic character. In a sane universe this would be called the quartic character. Let $p \in \mathbf{G}$ with $p \not\equiv 0$ modulo $1 + i$. Again, we do not assume p prime, so $F = \mathbf{G}/p\mathbf{G}$ need not be a field. The unit group, $U = \{1, -1, i, -i\}$, of G acts on $F - \{0\}$, splitting it into $\frac{N(p)-1}{4}$ orbits, each of cardinality 4, where the norm, $N(p)$, is defined by $N(p) = p\bar{p}$. We choose a set $R \subset \mathbf{G}$ of representatives for these orbits. If q is relatively prime to p and $r \in R$ then

$$qr \equiv \sigma_p(q, r) r' \quad \text{mod. } p \tag{2}$$

for a unique $\sigma_p(q, r) \in U$ and $r' \in R$. We define the *biquadratic character modulo p* by

$$\left\{\frac{q}{p}\right\} = \prod_{r \in R} \sigma_p(q, r). \tag{3}$$

This is independent of the choice of R, takes values in U, is multiplicative in q, and depends only on the residue of q modulo p, so we are justified in calling $\left\{\frac{q}{p}\right\}$ a biquadratic character modulo p.

If p is a prime then F is a finite field and $F - \{0\}$ a cyclic group of order $N(p) - 1$. In this case

$$q^{\frac{N(p)-1}{4}} \equiv \left\{\frac{q}{p}\right\} \quad \text{mod. } p \tag{4}$$

so $\left\{\frac{q}{p}\right\}$ is 1 if q is congruent to a fourth power, -1 if q is congruent to a square but not a fourth power, and $\pm i$ otherwise. If p is not a prime then (4) is usually false, so this provides another method for proving large numbers composite without exhibiting a factor.

We note that every $p \not\equiv 0$ modulo $1 + i$ satisfies the congruence

$$p \equiv \epsilon(p) \quad \text{mod. } 2 + 2i \tag{5}$$

for a unique $\epsilon(p) \in U$. We now have enough notation to state the main theorem about the biquadratic character.

THEOREM 36 (BIQUADRATIC RECIPROCITY)
If $p, q \in \mathbf{G}$ are relatively prime and neither is congruent to 0 modulo $1 + i$ then

$$\left\{\frac{p}{q}\right\} = (-1)^{\frac{N(p)-1}{4} \frac{N(q)-1}{4}} \frac{\epsilon(p)^{\frac{N(q)-1}{4}}}{\epsilon(q)^{\frac{N(p)-1}{4}}} \left\{\frac{q}{p}\right\}. \tag{6}$$

Before beginning the proof proper we need to establish some identities for the lemniscatic functions. The following are all immediate consequences of the corresponding identities for elliptic functions.

$$\mathrm{sl}(u + v)\,\mathrm{sl}(u - v) = \frac{(1 - \mathrm{sl}^2 u \, \mathrm{sl}^2 v)(\mathrm{sl}^2 u - \mathrm{sl}^2 v)}{(1 + \mathrm{sl}^2 u \, \mathrm{sl}^2 v)^2}, \tag{7}$$

$$sl(u + K) = \frac{cl\,u}{dl\,u}, \tag{8}$$

$$sl(u + 2K) = -sl\,u, \tag{9}$$

$$cl^2\,u = 1 - sl^2\,u, \tag{10}$$

$$dl^2\,u = 1 + sl^2\,u. \tag{11}$$

We derive other important identities from the integral representation. Recalling the proof of Theorem 32 we have the representation

$$f(z) = \int_0^z (1 - x^4)^{-1/2}\,dx \tag{12}$$

for the inverse function to sl near 0. The change of variable $y = ix$ gives the identity

$$f(iz) = i^{-1} f(z) \tag{13}$$

and hence

$$sl\,iu = i\,sl\,u. \tag{14}$$

From (10), (11) and (14) we deduce

$$cl\,iu = dl\,u \tag{15}$$

and

$$dl\,iu = cl\,u. \tag{16}$$

As in Theorem 32 the identities (14), (15), and (16), proved initially only for u sufficiently small, can be extended to all $u \in \mathbf{C}$, either by the rigidity theorem, or using the addition formulae. Since all elements of U are powers of i equation (14) implies

$$sl\,\epsilon u = \epsilon\,sl\,u \tag{17}$$

for all $\epsilon \in U$. From (9) and (14) we see that

$$sl(u + 2i\,K) = -sl\,u \tag{18}$$

and hence

$$\operatorname{sl}(u + (2 \pm 2i)K) = \operatorname{sl} u. \tag{19}$$

More generally

$$\operatorname{sl}(u + a\omega) = \operatorname{sl} u \tag{20}$$

for all $a \in \mathbf{G}$ where $\omega = (2 + 2i)K$. That these are the only periods is one of the exercises at the end of the section. Replacing v by iv in equation (7) and using (14) we find

$$\operatorname{sl}(u + iv)\operatorname{sl}(u - iv) = \frac{(1 + \operatorname{sl}^2 u \operatorname{sl}^2 v)(\operatorname{sl}^2 u + \operatorname{sl}^2 v)}{(1 - \operatorname{sl}^2 u \operatorname{sl}^2 v)^2}. \tag{21}$$

Multiplying (7) and (21) we obtain

$$\prod_{\epsilon \in U} \operatorname{sl}(u + \epsilon v) = \frac{\operatorname{sl}^4 u - \operatorname{sl}^4 v}{1 - \operatorname{sl}^4 u \operatorname{sl}^4 v}. \tag{22}$$

From the product representation we derive the multiplication formula

$$\operatorname{sl} qu = q \operatorname{sl} u \prod_s \frac{\operatorname{sl}(u + \frac{s}{q}\omega)}{\operatorname{sl} \frac{s}{q}\omega} \tag{23}$$

where q is not congruent to 0 modulo $1 + i$ and s ranges over over a set of representatives for $\mathbf{G}/p\mathbf{G}$ with 0 omitted. We use the identity (22) to write (23) as a product over a set S of representatives for the U-orbits on $\mathbf{G}/p\mathbf{G} - \{0\}$.

$$\operatorname{sl} qu = q \operatorname{sl} u \prod_{s \in S} \frac{\operatorname{sl}^4 \frac{s}{q}\omega - \operatorname{sl}^4 u}{(\operatorname{sl}^4 \frac{s}{q}\omega)(1 - \operatorname{sl}^4 \frac{s}{q}\omega) \operatorname{sl}^4 u}. \tag{24}$$

We can simplify (24) by evaluating the factor $\prod_{s \in S} \operatorname{sl}^4 \frac{s}{q}\omega$. The congruence (5) is simply a concise way to say that there exists an $a \in \mathbf{G}$ such that

$$q = \epsilon(q) + a(2 + 2i). \tag{25}$$

Thus

$$\operatorname{sl} qK = \operatorname{sl}(\epsilon(q)K + a\omega) \tag{26}$$

or, using equations (17) and (20),

$$\operatorname{sl} qK = \epsilon(q) \operatorname{sl} qK. \tag{27}$$

Taking $u = K$ in equation (24) and using (27), we find

$$\epsilon(q) = q \prod_{s \in S} \frac{\mathrm{sl}^4 \frac{s}{q}\omega - \mathrm{sl}^4 K}{(\mathrm{sl}^4 \frac{s}{q}\omega)(1 - \mathrm{sl}^4 \frac{s}{q}\omega\, \mathrm{sl}^4 K)}. \tag{28}$$

But equation (8) shows that $\mathrm{sl}^2 K = 1$, so

$$\epsilon(q) = q \prod_{s \in S} \frac{-1}{\mathrm{sl}^4 \frac{s}{q}\omega} \tag{29}$$

or

$$\prod_{s \in S} \mathrm{sl}^4 \frac{s}{q}\omega = (-1)^{\frac{N(q)-1}{4}} \frac{q}{\epsilon(q)}. \tag{30}$$

We can use (30) to rewrite (24) as

$$\mathrm{sl}\, qu = \epsilon(q) \prod_{s \in S} \frac{\mathrm{sl}^4 u - \mathrm{sl}^4 \frac{s}{q}\omega}{1 - \mathrm{sl}^4 u\, \mathrm{sl}^4 \frac{s}{q}\omega}. \tag{31}$$

With these identities at our disposal we can now prove Theorem 36.

Proof.   The congruence (2) asserts the existence of an $a \in \mathbf{G}$ such that

$$qr = \sigma_p(q, r)r' + ap. \tag{32}$$

Thus, using (17) and (20),

$$\mathrm{sl}\, \frac{qr}{p}\omega = \sigma_p(q, r)\, \mathrm{sl}\, \frac{r'}{p}\omega. \tag{33}$$

As r runs over R so does r'. Taking the product of (33) over all $r \in R$,

$$\prod_{r \in R} \mathrm{sl}\, \frac{qr}{p}\omega = \left\{\frac{q}{p}\right\} \prod_{r \in R} \mathrm{sl}\, \frac{r}{p}\omega. \tag{34}$$

This gives a formula

$$\left\{\frac{q}{p}\right\} = \prod_{r \in R} \frac{\mathrm{sl}\, \frac{qr}{p}\omega}{\mathrm{sl}\, \frac{r}{p}\omega} \tag{35}$$

for the biquadratic character in terms of the lemniscatic functions. We use the identity (31) to convert this to the more useful form

$$\left\{\frac{q}{p}\right\} = \epsilon(q)^{\frac{N(p)-1}{4}} \prod_{r \in R, s \in S} \frac{\mathrm{sl}^4 \frac{r}{p}\omega - \mathrm{sl}^4 \frac{s}{q}\omega}{1 - \mathrm{sl}^4 \frac{r}{p}\omega\, \mathrm{sl}^4 \frac{s}{q}\omega}. \tag{36}$$

Reversing the roles of p an q,

$$\left\{\frac{p}{q}\right\} = \epsilon(p)^{\frac{N(q)-1}{4}} \prod_{r \in R, s \in S} \frac{\mathrm{sl}^4 \frac{s}{q}\omega - \mathrm{sl}^4 \frac{r}{p}\omega}{1 - \mathrm{sl}^4 \frac{s}{q}\omega \, \mathrm{sl}^4 \frac{r}{p}\omega}. \tag{37}$$

Dividing equation (37) by (36) we find

$$\frac{\left\{\frac{p}{q}\right\}}{\left\{\frac{q}{p}\right\}} = \frac{\epsilon(p)^{\frac{N(q)-1}{4}}}{\epsilon(q)^{\frac{N(p)-1}{4}}} \prod_{r \in R, s \in S} (-1). \tag{38}$$

Since there are $\frac{N(p)-1}{4}\frac{N(q)-1}{4}$ factors in the product on the right this just gives equation (6), and the proof of the theorem is complete.

Exercises

1. Show that the periods given in equation (20) are the only periods of sl.

A Quick Review of Real Analysis

The purpose of this appendix is to provide the reader with exactly enough real analysis to read the remainder of the text. It is intended as a reference rather than as a substitute for a real analysis course. No attempt has been made at either elegance or generality.

The reader is assumed to be familiar with the differential and integral calculus. A good reference for this material is [12], by Courant and John. This material will normally be used without comment in the text. This appendix will cover a few of the more advanced topics.

In several places in the text we use Landau's "big O, little O" notation for the order of magnitude of functions. Most readers will be familiar with this already. Those who are not should consult pages 253-255 of [12].

Here we are concerned only with the analysis of continuous functions on subsets of $\mathbf{R}^n$ or $\mathbf{C}^n$. In what follows all sets are subsets of $\mathbf{R}^n$ or $\mathbf{C}^n$. A set U is called *open* if, for all $x \in U$, there is an $r > 0$ such that $B(x; r) \subset U$ where $B(x; r)$ is the ball of radius r about x,[11]

$$B(x; r) = \{y : |x - y| < r\}.$$

A *neighborhood* of a point is just an open set containing the point. A *punctured neighborhood* of a point z is just a set of the form $U - \{z\}$, where U is a neighborhood of z. A set is called *closed* if its complement is open. A set K is called *bounded* if there is some $r > 0$ such that $K \subset B(0; r)$. A set U is called *connected* if it cannot be written as the union of two disjoint, nonempty, open sets. The most important property of connected sets is that a function which is locally constant[12] is necessarily constant.

We will not, in this book, have occasion to integrate any function which is not continuous. It therefore matters very little whether we interpret integrals as

---

[11] This is the *open ball* of radius r about x. The *closed ball* is defined in the same way, except that we replace $|x - y| < r$ by $|x - y| \le r$.

[12] We say a function is locally X if the function is X in a neighborhood of every point. Thus a function is locally constant if the function is constant in a neighborhood of every point. Similar remarks apply to locally bounded, locally integrable, etc.

Riemann integrals or Lebesgue integrals.[13] The meaning of

$$\int_{[a,b]} f(x)\,dx$$

where f is continuous on the closed interval $[a, b]$ is entirely unambiguous. The integral

$$\int_{(a,b)} f(x)\,dx$$

over an open interval,[14] often called an *improper integral*, is

$$\int_{(a,b)} f(x)\,dx = \lim_{\substack{\alpha \to a^+ \\ \beta \to b^-}} \int_{[\alpha,\beta]} f(x)\,dx. \tag{1}$$

Here we merely assume that f is continuous on the open interval (a, b).[15] We say the integral is convergent or divergent according to whether this limit is convergent or divergent. Sometimes it is possible to prove convergence by Cauchy's criterion, Theorem K below. For this we must show that there are, for any $\epsilon > 0$, numbers $A(\epsilon)$ and $B(\epsilon)$ in the interval (a, b) such that

$$\left| \int_{[\alpha',\alpha'']} f(x)\,dx \right| \le \epsilon \tag{2}$$

---

[13] There is a common misperception that Lebesgue integration has somehow rendered Riemann integration obsolete. It is certainly true that any Riemann integrable function is Lebesgue integrable, while most Lebesgue integrable functions are not Riemann integrable. It is worth remembering, however, that theorems have conclusions as well as hypotheses. While it may be advantageous to replace Riemann integrability by Lebesgue integrability in the hypotheses of a theorem, the reverse is true of the conclusion. In some respects it is a matter of taste whether one prefers theorems which make weak statements under weak hypotheses or strong statements under strong hypotheses.

For some applications, though, one choice is often much more convenient. If one wants to use the techniques of elliptic partial differential equations, it is advantageous to use Lebesgue integration throughout. Since these are the most powerful techniques available in several complex variables it would be foolish to write a textbook on several complex variables based on the Riemann integral. In numerical analysis, by contrast, the Riemann integral is much more useful, since a Riemann sum can be evaluated in a finite amount of time. It would be unwise to base an algorithmic treatment of one complex variable on Lebesgue integration. For a book on continued fractions the optimal choice would probably be the Stieltjes integral. The only general rule is to use those definitions which are best adapted to the techniques one plans to use. For the subject matter of this book, functions of one complex variable, there is no strong reason to prefer either integral. Most of the material presented predates either definition of the integral and is relatively insensitive to the precise meaning attached to the integral sign.

[14] Possibly infinite or semi-infinite.

[15] If it happens that f is continuous on $[a, b]$ then of course $\int_{[a,b]} f(x)\,dx = \int_{(a,b)} f(x)\,dx$. We are not, of course claiming that the integral cannot be defined if f is not continuous, merely that we do not, for our present purposes, need to define it for discontinuous f.

for any $a < \alpha' \le \alpha'' \le A(\epsilon)$ and

$$\left| \int_{[\beta',\beta'']} f(x)\,dx \right| \le \epsilon \tag{3}$$

for any $B(\epsilon) \le \beta' \le \beta'' < b$.

We say the integral

$$\int_{(a,b)} f(x)\,dx$$

is *absolutely* convergent if the integral

$$\int_{(a,b)} |f(x)|\,dx$$

is convergent. In this case

$$\left| \int_{(a,b)} f(x)\,dx \right| \le \int_{(a,b)} |f(x)|\,dx. \tag{4}$$

More generally, If f and g are continuous on (a,b),

$$|f(x)| \le g(x) \tag{5}$$

and

$$\int_{(a,b)} g(x)\,dx$$

is convergent then

$$\int_{(a,b)} f(x)\,dx$$

is convergent and

$$\left| \int_{(a,b)} f(x)\,dx \right| \le \int_{(a,b)} g(x)\,dx. \tag{6}$$

In fact it suffices to check that

$$\int_{[\alpha,\beta]} g(x)\,dx$$

is bounded for $a < \alpha < \beta < b$ by a constant independent of α and β. If f depends on some additional parameter y and

$$|f(x,y)| \le g(x) \tag{7}$$

then the integral of f converges uniformly in y. This theorem is exactly analogous to the comparison theorem for infinite sums, Theorem B below, and its proof is nearly identical.

We will never actually write

$$\int_{(a,b)} f(x)\, dx$$

or

$$\int_{[a,b]} f(x)\, dx$$

in the text. We write

$$\int_a^b f(x)\, dx$$

and leave the reader to determine whether the integrand is defined and continuous on the closed interval $[a, b]$, in which case we mean the former, or not, in which case we mean the latter. This is slightly obnoxious, but traditional. In practice the burden is minor; it is usually clear which is meant, and usually irrelevant.

The following useful theorems are stated and proved in [12]. We provide in each case a reference to the volume, chapter, section and subsection in which they appear.

THEOREM A
In absolutely convergent series rearrangement of the terms does not affect the convergence, and the value of the sum of the series is unchanged, exactly as in finite sums. (I.7.1.c)

THEOREM B
If the terms of the series $\sum_{\nu=1}^{\infty} g_\nu(x)$ satisfy the condition $|g_\nu(x)| \le a_\nu$ where the numbers a_ν are positive constants which form a convergent series $\sum_{\nu=1}^{\infty} a_\nu$, then the series $\sum_{\nu=1}^{\infty} g_\nu(x)$ converges uniformly (and absolutely). (I.7.4.b)

THEOREM C
If a series of continuous terms converges uniformly on an interval, its sum is also a continuous function. (I.7.4.c)

We remark here that this might as well have been stated in terms of a sequence of continuous functions. In general, to convert a theorem about series into a theorem about sequences, we apply the theorem to the series

$$\sum_{n=0}^{\infty} a_n$$

where

$$a_n = \begin{cases} s_0 & \text{if } n = 0, \\ s_n - s_{n-1} & \text{if } n > 0. \end{cases} \tag{8}$$

This gives a theorem about the sequence s_n. To convert a theorem about sequences into one about series, we apply it to the sequence s_n where

$$s_n = \sum_{j=0}^{n} a_n. \tag{9}$$

This gives a theorem about the series

$$\sum_{n=0}^{\infty} a_n.$$

There is no difference between the theory of sequences and that of series and it is more economical to state only one of each pair of theorems. Similar remarks apply to infinite products.

THEOREM D
If in an interval the sequence of functions $f_1(x)$, $f_2(x)$, $\ldots$ tends uniformly to the limit function $f(x)$, then

$$\int_a^b f(x)\,dx = \lim_{n \to \infty} \int_a^b f_n(x)\,dx \tag{10}$$

for every pair of numbers a and b lying in the interval; in other words, we can interchange the order of operations of integration and passing to the limit. (I.7.4.d)

THEOREM E
If, on differentiating a convergent infinite series

$$\sum_{v=0}^{\infty} G_v(x) = F(x)$$

term by term, we obtain a uniformly convergent series of continuous terms

$$\sum_{v=0}^{\infty} g_v(x) = f(x),$$

then the sum of the last series is equal to the derivative of the sum of the first series. (I.7.4.e)

We note that the differentiability of F belongs to the conclusion rather than to the hypotheses of the theorem.

THEOREM F

If the series

$$\sum_{n=0}^{\infty} a_n$$

converges absolutely to A and the series

$$\sum_{n=0}^{\infty} b_n$$

converges absolutely to B, then the series

$$\sum_{n=0}^{\infty} c_n$$

converges absolutely to AB, where

$$c_n = a_0 b_n + a_1 b_{n-1} + \cdots + a_n b_0. \tag{11}$$

(I.7.A.1)

THEOREM G

The integral

$$F(x) = \int_a^b f(x, y) \, dy \tag{12}$$

is a continuous function of the parameter x, for $\alpha \le x \le \beta$ if $f(x, y)$ is continuous on the rectangle R given by $\alpha \le x \le \beta$, $a \le y \le b$. (II.1.8.b)

THEOREM H

If in the closed rectangle $\alpha \le x \le \beta$, $a \le y \le b$ the function $f(x, y)$ is continuous and has a continuous partial derivative with respect to x, we may differentiate the integral with respect to the parameter under the integral sign, that is

$$\frac{d}{dx} F(x) = \frac{d}{dx} \int_a^b f(x, y) \, dy = \int_a^b f_x(x, y) \, dy. \tag{13}$$

Moreover F' is a continuous function of x. (II.1.8.b)

We note that the differentiability of F belongs to the conclusion of the theorem rather than its hypotheses.

THEOREM I

Let $f(x, y)$ be continuous on the rectangle R given by $\alpha \leq x \leq \beta, a \leq y \leq b$. Then the integrals

$$I = \int_a^b d\xi \int_\alpha^\beta f(\xi, \eta) \, d\eta \qquad (14)$$

and

$$J = \int_\alpha^\beta d\eta \int_a^b f(\xi, \eta) \, d\xi \qquad (15)$$

have the same value.

THEOREM J

Every bounded infinite sequence of points $P_1, P_2, \ldots$ has a convergent subsequence. (II.1.A.1.b)

THEOREM K

A sequence of points $P_1, P_2, \ldots$ in the plane (and similarly a sequence in n-dimensional euclidean space) converges to a limit if and only if for every $\epsilon > 0$ there exists a number $N = N(\epsilon)$ such that the distance between P_n and P_m is less than ϵ whenever both n and m are greater than n. (II.1.A.1.b)

THEOREM L

If a function F is defined and continuous in a closed and bounded set, then it assumes a greatest value ("maximum") and a least value ("minimum") in S. (II.1.A.2)

We will only need this last theorem in the special case where S is a rectangle.

Next we make some remarks concerning Taylor's theorem. Taylor knew Taylor's theorem in the form

$$f(x) = \sum_{k=0}^\infty \frac{f^k(a)}{k!} (x - a)^k \qquad (16)$$

and seems to have regarded it as valid for all interesting functions f, see [66]. Theorem 16 in Section 2.4 gives the precise conditions under which (16) is valid. What is now called Taylor's theorem is the corresponding statement with the sum truncated at some exponent m. We follow this usage despite its historical inaccuracy. With this convention, there are two main forms of Taylor's theorem. It is only a minor exaggeration to describe them as the right form and the wrong form.

The wrong form, due to Lagrange [48], says that if f is m times continuously differentiable on the interval $[a, x]$ then

$$f(x) = \sum_{k=0}^{m-1} \frac{f^k(a)}{k!}(x-a)^k + \frac{f^m(t)}{k!}(x-a)^m \qquad (17)$$

for some t in $[a, x]$. The right form, whose origin is more obscure, says that, under the same hypotheses on f,

$$f(x) = \sum_{k=0}^{m-1} \frac{f^k(a)}{k!}(x-a)^k + \frac{1}{(m-1)!}\int_a^x (x-t)^{m-1} f^m(t)\, dt. \qquad (18)$$

Both are proved in [12], I.5.

It is easy to deduce (17) from (18), but not conversely. The proof of (18) is by induction; one gets from the formula for m to that for $m+1$ by integration by parts. One reason to call (17) the wrong form and (18) the right form is that the latter is true for complex valued f, while the former is false. Another reason is that nearly all the functions we consider depend on one or more additional parameters. The t in (17) is, in general, a rather poorly behaved function of these parameters. The dependence of the integral in (18) is generally much simpler. As an example we consider the smoothness[16] of the difference quotient of a smooth function,

$$g(x, y) = \frac{f(x) - f(y)}{x - y} \qquad (19)$$

Using the case $m = 0$ of (18) we see that

$$g(x, y) = \frac{\int_y^x f'(z)\, dz}{x - y} = \int_0^1 f'(tx + (1-t)y)\, dt. \qquad (20)$$

We can then use Theorem H to differentiate under the integral sign. To prove the same result using the form (17) of Taylor's theorem is much harder.[17]

We will occasionally need an inequality which follows easily from either form of Taylor's theorem,

$$e^x \geq 1 + x. \qquad (21)$$

This implies

$$e^x > x. \qquad (22)$$

---

[16]We say that a function is smooth if it has continuous derivatives of all orders.

[17]Of course equation (20) also follows from the fundamental theorem of calculus. This illustrates the crucial difference between (17) and (18). The latter generalizes the fundamental theorem of calculus while the former generalizes the mean value theorem. The fundamental theorem of calculus generalizes much better to several variables than the mean value theorem.

Taking $x = \epsilon \log t$, with $\epsilon > 0$, we find

$$\log t < \frac{t^\epsilon}{\epsilon}. \tag{23}$$

We now state some minor extensions of the preceding theorems, mostly concerned with open intervals and improper integrals. Most of these are simply the theorems already stated above, coupled with simple theorems about limits. A few rely on the fact, often used but rarely stated explicitly, that a locally continuous function is continuous. A simple consequence is the following extension of Theorem C.

THEOREM M

Let f_n be a sequence of continuous functions on an open set Ω. Suppose that for every $x \in \Omega$ there is an $r > 0$ such that the sequence f_n is uniformly convergent on the ball of radius r about x. Then its limit is a continuous function on Ω.

We call such a sequence locally uniformly convergent. There is a similar criterion for series, and also for limits as a continuous parameter tends to some limiting value. Certainly the hypotheses of the theorem are satisfied if f_n converges uniformly on every closed ball contained in Ω. If $\Omega \subset \mathbf{R}$, as it will be in most of our applications, then the closed balls contained in Ω are just the closed intervals contained in Ω, and we obtain a particularly simple criterion for the continuity of the limit.

The following theorem provides a useful analogue of Theorem D.

THEOREM N

If in an interval the sequence of functions $f_1(x)$, $f_2(x)$, ... tends locally uniformly to the limit function $f(x)$, then

$$\int_a^b f(x)\,dx = \lim_{n \to \infty} \int_a^b f_n(x)\,dx \tag{24}$$

for every pair of numbers a and b lying in the interval; in other words, we can interchange the order of operations of integration and passing to the limit.

We note that the interval is not assumed to be closed, but the points a and b must lie inside the interval. If the interval is open then a and b are not allowed to be the endpoints of the interval. This theorem follows immediately from Theorem D. We simply apply Theorem D not to the whole interval but only to the subinterval $[a, b]$. On this interval the sequence is uniformly convergent by hypothesis.

Theorems about integration and differentiation usually come in pairs, the link being the fundamental theorem of calculus. Theorems D and E are one such pair. Theorems G and H form another. The other member of the pair containing Theorem N is

THEOREM O

If, on differentiating a convergent infinite series

$$\sum_{v=0}^{\infty} G_v(x) = F(x)$$

term by term, we obtain a locally uniformly convergent series of continuous terms

$$\sum_{v=0}^{\infty} g_v(x) = f(x)$$

then the sum of the last series is equal to the derivative of the sum of the first series.

Once again the differentiability of F belongs to the conclusion rather than to the hypotheses of the theorem.

The most delicate operations in analysis are those involving the interchange of limits. The standard example

$$a_{m,n} = \frac{m}{m+n} \tag{25}$$

shows that it is entirely possible that each of the four limits

$$\lim_{m \to \infty} a_{m,n}$$

$$\lim_{n \to \infty} a_{m,n}$$

$$\lim_{m \to \infty} \lim_{n \to \infty} a_{m,n}$$

$$\lim_{n \to \infty} \lim_{m \to \infty} a_{m,n}$$

exists, but

$$\lim_{m \to \infty} \lim_{n \to \infty} a_{m,n} \neq \lim_{n \to \infty} \lim_{m \to \infty} a_{m,n}. \tag{26}$$

This example is somewhat artificial. A naturally occurring example may be found in Section 3.2. The following theorem gives conditions under which this behavior can be excluded.

THEOREM P

Suppose

1.

$$\lim_{m \to \infty} a_{m,n}$$

converges.

2.

$$\lim_{n \to \infty} a_{m,n}$$

converges uniformly *in n.*

3.

$$\lim_{m \to \infty} \lim_{n \to \infty} a_{m,n}$$

converges.

Then

$$\lim_{n \to \infty} \lim_{m \to \infty} a_{m,n}$$

converges and

$$\lim_{m \to \infty} \lim_{n \to \infty} a_{m,n} = \lim_{n \to \infty} \lim_{m \to \infty} a_{m,n}. \qquad (27)$$

The proof of this theorem is quite easy. We must check that, for any $\epsilon > 0$, there is an N such that

$$\left| \lim_{m \to \infty} \lim_{n \to \infty} a_{m,n} - \lim_{m \to \infty} a_{m,v} \right| < \epsilon \qquad (28)$$

whenever $v > N$. If $\epsilon > 0$ is given then, by hypothesis 2, there is an N such that

$$\left| a_{\mu,v} - \lim_{n \to \infty} a_{\mu,n} \right| < \frac{\epsilon}{3} \qquad (29)$$

if $v > N$. This N depends only on ϵ. There is also, by hypothesis 1, an M_v such that

$$\left| a_{\mu,v} - \lim_{m \to \infty} a_{m,v} \right| < \frac{\epsilon}{3} \qquad (30)$$

if $\mu > M_v$. This M_v might, in principle, depend on v as well as ϵ. Finally there is, by hypothesis 3, an M such that

$$\left| \lim_{m \to \infty} \lim_{n \to \infty} a_{m,n} - \lim_{n \to \infty} a_{\mu,n} \right| < \frac{\epsilon}{3}. \qquad (31)$$

This M depends only on ϵ. For any $\nu > N$ we can choose a $\mu > \max(M, M_\nu)$ and combine the three estimates above to give (28).

It is clear from the proof that the theorem remains true if one or both of the limits is with respect to a continuous parameter rather than a discrete parameter. The limit with respect to the continuous parameter may be either one-sided or two-sided. We obtain four new theorems by combining Theorem P with theorems D, G, H, and I.

Theorem N, as already noted, does not permit the interchange of limits and improper integrals, even in the case of local uniform convergence. It is not true for an arbitrary locally uniformly convergent sequence of functions $f_1(x), f_2(x), \ldots$ integrable on an open interval (a, b) that

$$\int_a^b f(x)\,dx = \lim_{n \to \infty} \int_a^b f_n(x)\,dx. \tag{32}$$

In practice this is less of a nuisance than it might appear. Before we can even attach a meaning to the right hand side of this equation, we must prove the existence of the limits

$$\lim_{\substack{\alpha \to a^+ \\ \beta \to b^-}} \int_{[\alpha, \beta]} f_n(x)\,dx$$

for each n. The following theorem shows that it suffices if these limits converge uniformly in n.

THEOREM Q
If in an open interval (a, b) the sequence of functions $f_1(x), f_2(x), \ldots$ tends locally uniformly to the limit function $f(x)$, if the limits

$$\lim_{\substack{\alpha \to a^+ \\ \beta \to b^-}} \int_{[\alpha, \beta]} f_n(x)\,dx$$

converge uniformly in n, and if the limit

$$\lim_{n \to \infty} \int_a^b f_n(x)\,dx$$

converges then

$$\int_a^b f(x)\,dx = \lim_{n \to \infty} \int_a^b f_n(x)\,dx. \tag{33}$$

In other words, we can interchange the order of operations of integration and passing to the limit.

THEOREM R
The integral

$$F(x) = \int_a^b f(x, y)\, dy \qquad (34)$$

is a continuous function of the parameter x, for $\alpha < x < \beta$ if $f(x, y)$ is continuous on the rectangle R given by $\alpha < x < \beta$, $a < y < b$ and the integral defining F converges locally uniformly in x.

THEOREM S
If in the open rectangle $\alpha < x < \beta$, $a < y < b$ the function $f(x, y)$ is continuous and has a continuous partial derivative with respect to x, we may differentiate the integral with respect to the parameter under the integral sign, that is

$$\frac{d}{dx} F(x) = \frac{d}{dx} \int_a^b f(x, y)\, dy = \int_a^b \frac{\partial f}{\partial x}(x, y)\, dy, \qquad (35)$$

provided that the integral on the right is convergent locally uniformly in x. Moreover F' is a continuous function of x.

We note again that the differentiability of F belongs to the conclusion of the theorem rather than to its hypotheses.

THEOREM T
Let $f(x, y)$ be continuous on the rectangle R given by $\alpha < x < \beta$, $a < y < b$. Then the integrals

$$I = \int_a^b d\xi \int_\alpha^\beta f(\xi, \eta)\, d\eta \qquad (36)$$

and

$$J = \int_\alpha^\beta d\eta \int_a^b f(\xi, \eta)\, d\xi \qquad (37)$$

have the same value, provided that all the integrals appearing are convergent and that one of the inner integrals is convergent locally uniformly.

We need some definitions and one theorem about line integrals in $\mathbf{R}^2$. Let Ω be an open subset of $\mathbf{R}^2$. A *path* in Ω is a pair (x, y) of twice continuously differentiable functions on the interval $[0, 1]$ such that

$$(x, y)(t) \in \Omega \qquad (38)$$

for $0 \leq t \leq 1$. The path is called *closed* if

$$(x, y)(1) = (x, y)(0). \tag{39}$$

A *homotopy* between closed paths (x', y') and (x'', y'') is a pair (ξ, η) of twice continuously differentiable functions on the square $[0, 1] \times [0, 1]$ such that

$$(\xi, \eta)(s, t) \in \Omega \tag{40}$$

for all $0 \leq s, t \leq 1$,

$$(\xi, \eta)(s, 1) = (\xi, \eta)(s, 0) \tag{41}$$

for all $0 \leq s \leq 1$,

$$(\xi, \eta)(0, t) = (x', y')(t) \tag{42}$$

for all $0 \leq t \leq 1$, and

$$(\xi, \eta)(1, t) = (x'', y'')(t) \tag{43}$$

for all $0 \leq t \leq 1$. Two paths are called *homotopic* if there is a homotopy between them. A homotopy is most naturally thought of as a family of paths. If we define

$$(x_s, y_s)(t) = (\xi, \eta)(s, t) \tag{44}$$

then equation (40) says that (x_s, y_s) is a path in Ω for all $0 \leq s \leq 1$, equation (41) says this path is closed, and equations (42) and (43) say that (x_s, y_s) is (x', y') for $s = 0$ and (x'', y'') for $s = 1$. A path is called *contractible* if it is homotopic to a constant path. If (M, N) is a pair of continuous functions on Ω then the *line integral* is defined by

$$\int (M \, dx + N \, dy) = \int_0^1 \left[M(x(t), y(t)) \frac{dx}{dt}(t) + N(x(t), y(t)) \frac{dy}{dt}(t) \right] dt. \tag{45}$$

It is clear from this definition that the line integral along a constant path is zero. The only fact about line integrals which we need in the text is the following theorem.

THEOREM U
Let Ω be an open subset of $\mathbf{R}^2$ and (M, N) a pair of continuously differentiable functions on Ω. The following conditions are equivalent.

   1. $\int (M \, dx' + N \, dy') = \int (M \, dx'' + N \, dy'')$ for any homotopic pair of closed paths (x', y') and (x'', y'').

2. $\int (M\,dx + N\,dy) = 0$ *for any closed contractible path* (x, y).

3. $(\frac{\partial N}{\partial x} - \frac{\partial M}{\partial y})(\overline{x}, \overline{y}) = 0$ *for any* $(\overline{x}, \overline{y}) \in \Omega$.

We give a quick proof of the theorem, the heart of which is the following calculation. Let (ξ, η) be a homotopy between closed paths (x', y') and (x', y'). We define a pair (P, Q) of functions on the square $[0, 1] \times [0, 1]$ by

$$P(s, t) = M(\xi(s, t), \eta(s, t))\frac{\partial \xi}{\partial s}(s, t) + N(\xi(s, t), \eta(s, t))\frac{\partial \eta}{\partial s}(s, t) \qquad (46)$$

$$Q(s, t) = M(\xi(s, t), \eta(s, t))\frac{\partial \xi}{\partial t}(s, t) + N(\xi(s, t), \eta(s, t))\frac{\partial \eta}{\partial t}(s, t). \qquad (47)$$

Equations (42) and (47) show that

$$\int (M\,dx' + N\,dy') = \int_0^1 Q(0, t)\,dt, \qquad (48)$$

while equations (43) and (47) show that

$$\int (M\,dx'' + N\,dy'') = \int_0^1 Q(1, t)\,dt. \qquad (49)$$

Subtracting equation (49) from equation (48) and using the fundamental theorem of calculus we find

$$\int (M\,dx' + N\,dy') - \int (M\,dx'' + N\,dy'') = \int_0^1 \int_0^1 \frac{\partial Q}{\partial s}(s, t)\,ds\,dt. \qquad (50)$$

Equation (41) implies

$$P(s, 0) = P(s, 1) \qquad (51)$$

and thus, by the fundamental theorem of calculus and Theorem I,

$$0 = \int_0^1 \int_0^1 \frac{\partial P}{\partial t}(s, t)\,ds\,dt. \qquad (52)$$

Subtracting (52) from (50) we find

$$\int (M\,dx' + N\,dy') - \int (M\,dx'' + N\,dy'') = \int_0^1 \int_0^1 \left(\frac{\partial Q}{\partial s} - \frac{\partial P}{\partial t}\right)(s, t)\,ds\,dt. \qquad (53)$$

A simple calculation gives the following formula for the integrand,

$$\left(\frac{\partial Q}{\partial s} - \frac{\partial P}{\partial t}\right)(s, t) = \left[\left(\frac{\partial N}{\partial x} - \frac{\partial M}{\partial y}\right)(\xi(s, t), \eta(s, t))\right] \tag{54}$$

$$\times \left[\left(\frac{\partial \xi}{\partial s}\frac{\partial \eta}{\partial t} - \frac{\partial \xi}{\partial t}\frac{\partial \eta}{\partial s}\right)(s, t)\right].$$

Now we proceed to the proof of the theorem. We will prove $1 \implies 2 \implies 3 \implies 1$. Suppose then that condition 1 is satisfied. If (x, y) is a contractible path then it is, by definition, homotopic to a constant path. By condition 1 the line integral along (x, y) is the same as the line integral along this constant path. But we have already observed that the line integral along a constant path is zero. Therefore condition 2 is satisfied.

Suppose now that condition 2 holds. The path

$$x'(t) = \bar{x} + r\cos(2\pi t), \tag{55}$$

$$y'(t) = \bar{y} + r\sin(2\pi t) \tag{56}$$

is contractible. Indeed,

$$\xi(s, t) = \bar{x} + sr\cos(2\pi t), \tag{57}$$

$$\eta(s, t) = \bar{y} + sr\sin(2\pi t) \tag{58}$$

is a homotopy to the constant path $(x'', y'')(t) = (\bar{x}, \bar{y})$. The line integral along the constant path (x'', y'') is zero so condition 2 implies

$$\int (M\,dx' + N\,dy') - \int (M\,dx'' + N\,dy'') = \int (M\,dx' + N\,dy') = 0. \tag{59}$$

Equation (53) shows that

$$\int_0^1 \int_0^1 \left(\frac{\partial Q}{\partial s} - \frac{\partial P}{\partial t}\right)(s, t)\,ds\,dt = 0 \tag{60}$$

or, using equation (54) and dividing by πr^2,

$$2\int_0^1 \int_0^1 \left(\frac{\partial N}{\partial x} - \frac{\partial M}{\partial y}\right)(\xi(s, t), \eta(s, t))s\,ds\,dt = 0. \tag{61}$$

As r tends to zero, the integrand tends uniformly to

$$\left(\frac{\partial N}{\partial x} - \frac{\partial M}{\partial y}\right)(\bar{x}, \bar{y})$$

so Theorem D shows that

$$\left(\frac{\partial N}{\partial x} - \frac{\partial N}{\partial x}\right)(\bar{x}, \bar{y}) = 0. \tag{62}$$

Thus we see that condition 3 is satisfied.

Suppose now that condition 3 is satisfied. Equation (54) shows that

$$\left(\frac{\partial Q}{\partial s} - \frac{\partial P}{\partial t}\right)(s, t) = 0 \tag{63}$$

for all $0 \le s, t \le 1$. From equation (53) we conclude that

$$\int (M\,dx' + N\,dy') = \int (M\,dx'' + N\,dy'') \tag{64}$$

for any homotopic pair of closed paths (x', y') and (x'', y''). Thus condition 1 is satisfied, and the proof of the theorem is complete.

Next we gather some miscellaneous facts which are used in the text. A useful trick for infinite sums, analogous to integration by parts, is Abel's technique of summation by parts.

THEOREM V

If

$$\lim_{n \to \infty} a_n b_n$$

converges and one of the two series

$$\sum_{n=1}^{\infty} a_n (b_{n+1} - b_n).$$

or

$$\sum_{n=2}^{\infty} (a_n - a_{n-1}) b_n$$

converges then so does the other, and

$$\sum_{n=1}^{\infty} a_n (b_{n+1} - b_n) = \lim_{n \to \infty} a_n b_n - a_1 b_1 - \sum_{n=2}^{\infty} (a_n - a_{n-1}) b_n. \tag{65}$$

The proof is quite simple. One considers the partial sums

$$
\begin{aligned}
\sum_{n=1}^{N} a_n(b_{n+1} - b_n) &= \sum_{n=1}^{N} a_n b_{n+1} - \sum_{n=1}^{N} a_n b_n \\
&= \sum_{n=1}^{N} a_n b_{n+1} - \sum_{n=2}^{N+1} a_n b_n + a_{N+1} b_{N+1} - a_1 b_1 \\
&= \sum_{n=2}^{N+1} a_{n-1} b_n - \sum_{n=2}^{N+1} a_n b_n + a_{N+1} b_{N+1} - a_1 b_1 \\
&= a_{N+1} b_{N+1} - a_1 b_1 - \sum_{n=2}^{N+1} (a_n - a_{n-1}) b_n .
\end{aligned}
\tag{66}
$$

and then lets N tend to infinity. An important special case occurs when $a_n = 1$ for all n and

$$
\lim_{n \to \infty} b_n = 0.
\tag{67}
$$

In this case we conclude that

$$
\sum_{n=1}^{\infty} (b_{n+1} - b_n) = -b_1 .
\tag{68}
$$

The most important estimate on integrals is that which says that the integral of a nonnegative function is nonnegative. All other inequalities are ultimately derived from this one. An easy consequence is

$$
\sup_{s \in I} \int_J h(s, t) \, dy \le \int_J \sup_{s \in I} h(s, t) \, dt.
\tag{69}
$$

Reversing the signs we find

$$
\int_J \inf_{s \in I} h(s, t) \, dt \le \inf_{s \in I} \int_J h(s, t) \, dt.
\tag{70}
$$

This can be used to prove various other inequalities. The main inequality we will need is Hölder's inequality.

THEOREM W
Let J be an interval and $1 < p, q < \infty$ real numbers with

$$
\frac{1}{p} + \frac{1}{q} = 1.
\tag{71}
$$

Then

$$
\int_J |f(t)g(t)| \, dt \le \left(\int_I |f(x)|^p \, dx \right)^{1/p} \left(\int_J |g(t)|^q \, dt \right)^{1/q}.
\tag{72}
$$

For our purposes it is more convenient to prove the equivalent from

$$\int_J |f(t)g(t)|\, dt \le \left(\int_I |f(x)|^{1/\alpha}\, dx \right)^{\alpha} \left(\int_J |g(t)|^{1/\beta}\, dt \right)^{\beta} \tag{73}$$

for any $0 < \alpha, \beta < 1$ with $\alpha + \beta = 1$. We begin by noting that, for $u, v > 0$.

$$\min_{0 < s < \infty} (\alpha s^{1/\alpha} u^{1/\alpha} + \beta s^{-1/\beta} v^{1/\beta}) = uv. \tag{74}$$

This can be proved by the standard method of the differential calculus. The derivative with respect to s of the expression on the right hand side vanishes only at

$$s = u^{-\beta} v^{\alpha}. \tag{75}$$

For larger values of s this derivative is positive; for smaller values it is negative. Thus this value of s gives the minimum. We can weaken the hypotheses to

$$u, v \ge 0 \tag{76}$$

if we replace (74) by

$$\inf_{0 < s < \infty} (\alpha s^{1/\alpha} u^{1/\alpha} + \beta s^{-1/\beta} v^{1/\beta}) = uv. \tag{77}$$

Now set

$$h(s, t) = (\alpha s^{1/\alpha} |f(t)|^{1/\alpha} + \beta s^{-1/\beta} |g(t)|^{1/\beta}). \tag{78}$$

From (77) we see that

$$\int_J \inf_{s \in I} h(s, t)\, dt = \int_J |f(t)g(t)|\, dt \tag{79}$$

with $I = (0, \infty)$, while

$$\inf_{s \in I} \int_J h(s, t)\, dt = \left(\int_I |f(x)|^{1/\alpha}\, dx \right)^{\alpha} \left(\int_J |g(t)|^{1/\beta}\, dt \right)^{\beta}. \tag{80}$$

The inequality (73) then follows immediately from (70).

We now consider the error committed in approximating an integral by the value of the integrand at the midpoint. If f is twice continuously differentiable on $[x - h, x + h]$ then

$$\left| \frac{1}{2h} \int_{x-h}^{x+h} f(u)\, du \right| \le \frac{h^2}{6} \max_{[x-h, x+h]} |f''|. \tag{81}$$

To prove this we calculate

$$\frac{1}{2h} \int_{x-h}^{x+h} f(u)\, du$$

by integration by parts. First we split the integral,

$$\frac{1}{2h} \int_{x-h}^{x+h} f(u)\, du = \frac{1}{2h} \int_{x-h}^{x} f(u)\, du + \frac{1}{2h} \int_{x}^{x+h} f(u)\, du. \qquad (82)$$

Next we integrate both integrals on the right by parts, choosing the constants of integration such that the boundary terms at $x \pm h$ vanish,

$$\frac{1}{2h} \int_{x-h}^{x+h} f(u)\, du = f(0) - \frac{1}{2h} \int_{x-h}^{x} (u - x + h) f'(u)\, du \qquad (83)$$

$$- \frac{1}{2h} \int_{x}^{x+h} (u - x - h) f'(u)\, du.$$

Then we integrate by parts again, still choosing the constants of integration such that the boundary terms at $x \pm h$ vanish. This has the pleasant side effect that the boundary terms at x in the two integrals cancel,

$$\frac{1}{2h} \int_{x-h}^{x+h} f(u)\, du = f(0) + \frac{1}{4h} \int_{x-h}^{x} (u - x + h)^2 f''(u)\, du \qquad (84)$$

$$+ \frac{1}{4h} \int_{x}^{x+h} (u - x - h)^2 f''(u)\, du.$$

Finally we estimate $f''(u)$ by its maximum on the domain of integration.

A similar result,

$$\left| \frac{f(x + h) - f(x)}{h} - f'(x) \right| \leq \frac{h}{2} \max_{x \leq \xi \leq x+h} |f''(\xi)|, \qquad (85)$$

holds for the approximation of a derivative by a difference quotient. It is easy to derive this by integration by parts, but this is unnecessary. Instead we apply Taylor's theorem with $m = 2$,

$$f(x + h) = f(x) + f'(x)h + \int_{x}^{x+h} (x + h - \xi) f''(\xi)\, d\xi \qquad (86)$$

or

$$\frac{f(x + h) - f(x)}{h} = f'(x) + h^{-1} \int_{x}^{x+h} (x + h - \xi) f''(\xi)\, d\xi. \qquad (87)$$

Estimating $f''(u)$ by its maximum on the domain of integration gives precisely the inequality (81).

Bibliography

Acta Math.	Acta Mathematica
Ann Math.	Annals of Mathematics
Ann. SSB	Annales de la Societe Scientifique de Bruxelles
Berl. A.	Abhandlungen der Königlicher Akademie der Wissenschaften zu Berlin
Berl. M.	Monatsberichte der Königlicher Akademie der Wissenschaften zu Berlin
Berl. S.	Sitzungsberichte der Königlicher Akademie der Wissenschaften zu Berlin
Bull. AMS	Bulletin of the American Mathematical Society
Bull. SMF	Bulletin de la Societeé Mathematique de France
Cam. Dub. Math. J.	Cambridge and Dublin Mathematical Journal
Comm. ASP	Commentarii Academiae Imperialis Scientiarum Petropolitanae
Comm. Gottg. rec.	Commentarii Gottingenses Recentiores
Crelle	Journal für die Reine und Angewandte Mathematik
Gött. A.	Abhandlungen der Königlicher Akademie der Wissenschaften zu Götingen
J. Ecole Poly.	Journal de l'Ecole Polytechnique
Mem. AIS	Mémoires de l'Académie Impériale des Sciences St. Pétersbourg
Mem. AS	Mémoires de l'Académie des Sciences
Proc. LMS	Proceedings of the London Mathematical Society
Proc. NAS	Proceedings of the National Academy of Sciences

[1] N. H. Abel. Recherches sur les fonctions elliptiques. *Crelle*, 2:101–181, 1827.

[2] E. W. Barnes. A new development of the theory of hypergeometric functions. *Proc. LMS*, 6:141–177, 1908.

[3] F. W. Bessel. Untersuchung des Theils der planetarischen Störungen, welcher aus der Bewegung der Sonne entsteht. *Berl. A.*, 1824.

[4] C. W. Borchardt. Theorie der elliptischen Funktionen aus den Eigenschaften der Thetareihen abgeleitet. 1880. Included in [41].

[5] H. Buchholz. *Die konfluente hypergeometrische Funktion.* Springer, Berlin, 1953.

[6] M. L. Cartwright. *Integral Functions.* University Press, Cambridge, 1956.

[7] A.-L. Cauchy. Sur un nouveau genre de calcul analogue au calcul infinitésimal. In *Exercices de mathématique*, page 26. De Bure, Paris, 1826. Reprinted in [8], Series II, Tome 6.

[8] A.-L. Cauchy. *Oeuvres*. Gauthier-Villars, Paris, 1882.

[9] P. L. Chebyshev. Sur la totalité des nombres premiers inférieure à une limite donnée. *Mem. AIS*, 6:141–157, 1851. Reprinted in [11].

[10] P. L. Chebyshev. Mémoire sur les nombres premiers. *Mem. AIS*, 7:13–33, 1854. Reprinted in [11].

[11] P. L. Chebyshev. *Oeuvres*. Académie Impériale des Sciences, St. Petersburg, 1899.

[12] R. Courant and F. John. *Introduction to Calculus and Analysis*. Wiley Interscience, New York, 1965.

[13] C. L. X. J. de la Vallée Poussin. Recherches analytiques sur la théorie des nombres. *Ann. SSB*, 20:183–256, 1896.

[14] A. de Moivre. Approximation ad summam terminorum binomii $(a + b)^n$ in seriem expansi. 1733. Distributed privately. An abridged translation appears in [15].

[15] A. de Moivre. *The Doctrine of Chances*. 2nd edition, 1738.

[16] P. G. Lejeune Dirichlet. Sur les intégrales Eulériennes. *Crelle*, 15:258–263, 1836. Reprinted in [18], vol. I, p. 271-277.

[17] P. G. Lejeune Dirichlet. Sur l'usage des séries infinies dans la théorie des nombres. *Crelle*, 18:259–274, 1838. Reprinted in [18].

[18] P. G. Lejeune Dirichlet. *Werke*. Reimer, Berlin, 1889.

[19] G. Doetsch. *Einführung in Theorie und Anwendung der Laplace-Transformation*. Birkhäuser, Basel, 1958.

[20] G. Eisenstein. Beiträge zur Theorie der elliptischen Funktionen. *Crelle*, 30:185–210, 1846. Reprinted in [21].

[21] G. Eisenstein. *Mathematische Abhandlungen*. Reimer, Berlin, 1847.

[22] A. Erdelyi, W. Magnus, and F. Oberhettinger. *Higher Transcendental Functions*. McGraw-Hill, New York, 1953.

[23] P. Erdős. On a new method in elementary number theory which leads to an elementary proof of the prime number theorem. *Proc. NAS*, 35:374–384, 1949.

[24] L. Euler. Variae observationes circa series infinitae. *Comm. ASP*, 9, 1737. Reprinted in [28].

[25] L. Euler. Methodus universalis series summandi ulterius promota. *Comm. ASP*, VIII:147–158, 1741. Reprinted in [28].

[26] L. Euler. *Introductio in analysin infinitorum*. Buosquet & Co., Lausanne, 1748. Reprinted in [28].

[27] L. Euler. *Institutiones calculi integralis*. Acad. Sci. Imp. Petrop., 1768. Reprinted in [28].

[28] L. Euler. *Opera Omnia*. Teubner, Leipzig, 1921.

[29] J. B. J. Fourier. *La Theorie Analytique de la Chaleur*. Firmin Didot, Paris, 1822.

[30] C. F. Gauss. *Demonstratio nova theorematis omnem functionem algebraicam rationalem integram unius variabilis in factores reali primi vel secundi gradus resolvi posse*. Fleckeisen, Helmstädt, 1799. Reprinted in [32].

[31] C. F. Gauss. Disquisitiones generales circa seriem infinitam. *Comm. Gottg. rec.*, II, 1813. Reprinted in [32], vol. III.

[32] C. F. Gauss. *Werke*. Springer, Berlin, 1863-1929.

[33] J. Hadamard. Sur la distribution des zéros de la fonction $\zeta(s)$ et ses conséquences arithmétiques. *Bull. SMF*, 24:199–220, 1896. Reprinted in [34].

[34] J. Hadamard. *Oeuvres*. CNRS, Paris, 1968.

[35] P. Henrici. *Applied and Computational Complex Analysis*. Wiley, New York, 1977.

[36] E. W. Hobson. *Theory of Spherical and Ellipsoidal Harmonics*. University Press, Cambridge, 1931.

[37] L. Hörmander. *Introduction to Complex Analysis in Several Variables*. van Nostrand, Princeton, 1966.

[38] C. G. J. Jacobi. *Fundamenta Nova Theoriae Functionum Ellipticarum*. Bornträger, Königsberg, 1829. Reprinted in [41].

[39] C. G. J. Jacobi. De usu legitimo formulae summatoriae MacLaurinianae. *Crelle*, 12:263–272, 1834. Reprinted in [41].

[40] C. G. J Jacobi. Untersuchungen über die Differentialgleichung der hypergeometrischen Reihe. *Crelle*, 56:149–165, 1859. Written perhaps in 1843, found among Jacobi's papers after his death and edited for publication by E. Heine, later reprinted in [41].

[41] C. G. J. Jacobi. *Mathematische Werke*. Reimer, Berlin, 1881.

[42] C.G.J. Jacobi. Über die partielle Differentialgleichung, welcher die Zähler und Nenner der elliptischen Functionen genüge leisten. *Crelle*, 36:81–88, 1848.

[43] L. Kronecker. Zur Theorie der elliptischen Funktionen XI. *Berl. S.*, 1886. Reprinted in [44].

[44] L. Kronecker. *Werke.* Teubner, Leipzig, 1895-1929.

[45] E. E. Kummer. Über die hypergeometrische Reihe $1 + \frac{\alpha \cdot \beta}{1 \cdot \gamma} x + \cdots$. *Crelle*, 15:39–83,127–172, 1836. Reprinted in [47]. Translated into English by H. Nagaoka and printed in vol. IV of the journal of the Tokyo Mathematical Society, pages 273-325, 535-405 (1891).

[46] E. E. Kummer. De integralibus quibusdam definitis et seriebus infinitis. *Crelle*, 17:228–242, 1837. Reprinted in [47].

[47] E. E. Kummer. *Collected Papers.* Springer, Berlin, 1975.

[48] J.-L. Lagrange. *Oeuvres.* Gauthier-Villars, Paris, 1867.

[49] A.-M. Legendre. Recherches sur la figure des planètes. *Mem. AS*, page 370ff, 1784.

[50] A. M. Legendre. *Essai sur la Theorie des Nombres.* Firmin Didot, Paris, 2nd edition, 1808.

[51] B. J. Levin. *Distribution of the Zeros of Entire Functions.* AMS, Providence, 1964.

[52] C. MacLaurin. *Treatise on Fluxions.* Ruddimans, Edinburgh, 1742.

[53] A. I. Markushevich. *Theory of Functions of a Complex Variable.* Chelsea, New York, 1965.

[54] Z. Nehari. *Conformal Mapping.* McGraw-Hill, New York, 1952.

[55] F. W. Newman. On Γa, especially when a is negative. *Cam. Dub. Math. J.*, 3, 1848.

[56] L. Olivier. Remarques sur les séries infinies et leur convergence. *Crelle*, 2:31–44, 1827.

[57] O. Perron. *Die Lehre von den Kettenbrüchen.* Teubner, Stuttgart, 3rd edition, 1957.

[58] H. Poincaré. Sur les intégrales irrégulières des équations linéaires. *Acta Math.*, 8:295–344, 1886. Reprinted in [59].

[59] H. Poincaré. *Oeuvres.* Gauthier-Villars, Paris, 1951.

[60] C. F. B. Riemann. Über die Anzahl der Primzahlen unter einer gegebenen Grenze. *Berl. M.*, pages 671–680, 1859.

[61] A. Selberg. An elementary proof of the prime number theorem. *Ann Math.*, 50:305–318, 1949.

[62] S. Skewes. On the difference $\pi(x) - \operatorname{li} x$. *Proc. LMS*, 5:48–70, 1955.

[63] T. J. Stieltjes. Recherches sur les fractions continues. *Ann. Toulouse*, 8:1–122, 1894.

[64] J. Stirling. *Methodus Differentialis*. William Bowyer, London, 1730.

[65] G. Szegő. *Orthogonal Polynomials*. AMS, New York, 1959.

[66] B. Taylor. *Methodus Incrementorum Directa et Inversa*. Pearson, London, 1717.

[67] L. W. Thomé. Zur Theorie der linearen Differentialgleichungen. *Crelle*, 87:222–349, 1879.

[68] E. C. Titchmarsh. *Introduction to the Theory of Fourier Integrals*. Clarendon, Oxford, 1937.

[69] J. Wallis. *Arithmetica Infinitorum*. Leon Lichfield (Oxford Unversity Press), Oxford, 1655. Reprinted in [70].

[70] J. Wallis. *Opera Mathematica*. Oxford University, Oxford, 1695.

[71] G. N. Watson. *Treatise on the Theory of Bessel Functions*. University Press, Cambridge, 2nd edition, 1944.

[72] H. Weber. *Lehrbuch der Algebra*, volume III. Friedrich Vieweg, Braunschweig, 2nd edition, 1908.

[73] K. Weierstrass. Uber die Theorie der analytischen facultäten. *Crelle*, 51, 1856. Reprinted in [74], vol. I, pp. 153-221.

[74] K. Weierstrass. *Mathematische Werke*. Mayer & Müller, Berlin, 1894.

[75] E. T. Whittaker and G. N. Watson. *A Course of Modern Analysis*. Cambridge University Press, Cambridge, 1927.

[76] E.T. Whittaker. An expression for certain known functions as generalized hypergeometric functions. *Bull. AMS*, 10:125–134, 1904.

Index